SCIENCE, PSYCHOLOGY, AND THE REALMS OF SPIRIT

Joyce A. Kovelman
Ph.D. Anatomy, Ph.D. Psychology

JALMAR PRESS
Carson, California

Once Upon ASOUL
The Story Continues. . .

Science, Psychology, and The Realms of Spirit

Copyright © 1998 by Joyce A. Kovelman, Ph.D. Anatomy, Ph.D. Psychology

All rights reserved. No part of this book may be reproduced by any mechanical, photographic, or electronic process, or in the form of a photographic recording, nor may it be stored in a retrieval system, transmitted or otherwise copied for public or private use without the written permission of the publisher. Requests for permission should be addressed to:

> Jalmar Press
> Permissions Dept.
> 24426 S. Main St., Suite 702
> Carson, CA 90745
> (310)816-3085 Fax (310)816-3092
> email: blwjalmar@worldnet.att.net

Library of Congress Cataloging-in-Publication Data

Kovelman, Joyce Ann.
    Once Upon ASOUL: The Story Continues . . . Science, Psychology, and The Realms of Spirit /
    by Joyce A. Kovelman.
       p.   cm.
    Companion volume to Namasté.
    Includes bibliographical references and index.
    ISBN 1-880396-52-1
    1. Spiritual life—Miscellanea. 2. Psychology—Miscellanea.
3. Science—Miscellanea.   I. Title.
BF1999.K695 1997
291.4—dc21   97-25570
                                                                       CIP

Published by Jalmar Press

25+YEARS
AWARD WINNING
PUBLISHER
𝒫

Once Upon ASOUL
The Story Continues. . .

    Author: Joyce A. Kovelman, Ph.D. Anatomy, Ph.D. Psychology
    Editor: Susan Remkus
    Project Director: Jeanne Iler
    Production
       & Design: Electronic Publishing Services, Inc.

Manufactured in the United States of America

10 9 8 7 6 5 4 3 2 1
ISBN: 1-880396-52-1

Dedicated to the Emerging Feminine

Once Upon A SOUL has been selected as a textbook for University of the Air. Dr. Kovelman's must-read book has profound implications for our further understanding of the interdisciplinary connections of science, psychology and spirit, as well as for human survival. *Once Upon A SOUL* is a beautiful work that will become a classic.

—F. Richard Schneider, Ph.D., Chancellor,
Institute of Global Education

Dr. Joyce Kovelman's book, *Once Upon A SOUL,* is a very impressive synthesis of the physical, psychological and spiritual forces which mold and control our daily lives. The book is heuristic at all levels from the overall thesis of man's evolutionary progress to as specific and limited as the theme of any single paragraph. This makes it possible for the reader to obtain help from a level of generalities to the specifics of a particular question. There are ample footnotes and appendices of suggested readings provided for us by Dr. Kovelman to delve as deeply as we are so inclined. This book is a truly fulfilling and profitable experience.

—Roy S. Griffiths, Ph.D., Professor Emeritus, Psychology,
California State University, Northridge

Acknowledgments

Heartfelt thanks to my editor and friend, Andrea Cagan, whose gentleness and support were invaluable. Special thanks to Susan Herner, who encouraged and guided my dream from the very beginning. Enormous love and appreciation to my "unofficial" editor, Sarah Minden, who listened and supported my daily unfolding. J.T. O'Hara's gifts of enthusiasm and humor were indispensible. Endless gratitude to Carolina Loren, for her steadfast belief in my ideas and visions.

Love and admiration for Susan Tereba, whose inspiration and vision guided my ideas. I am deeply moved by the encouragement and dedication of my publisher Dr. Bradley Winch, Sr. at Jalmar Press and the gentleness of my copy editor, Susan Remkus.

Without the loving support of my husband, Gilbert, this book would never have been possible. He willingly listened to my ideas, nightly. I am indebted to my parents, Sara and Victor, and to my children, Paul, Robert, Bruce, Gabrielle, Alexandra, Noreen, and Angela for their patience and encouragement. Many friends have also read portions of this book and generously offered their comments, insights, and suggestions. They are very much a part of this endeavor. Thank you Elizabeth Caulder, Alroy Flack, Kay Fox, Peter Gelfer, Roy Griffiths, Elizabeth Iwanski, Richard Schneider, and Mark Sudock.

There are so many people who have helped make this book possible. Throughout my journey, I have been blessed with many wonderful teachers. Special thanks to Peter DiCiaula, Ted Falcon, Peter Gelfer, Violet Hamilton, James Jacobs, Monroe Kline, Grandpa Koshansky, Ken McCloud, Peter Moscow, Lopsang Rapgay, Jane Roberts, Arnold B. Scheibel, and Uncle Ira. Your belief in me has deeply touched my heart. And my special gratitude to Alexandra, Mother Earth, and Father Sky.

My special thanks to Bryan and to Jeanne.

Preface

I was born an identical twin. During my childhood, it was my primary identity, although I devoted much of my time to defining myself as "other than." My mother adored having twins, and so we were raised as a pair, rather than as two individuals. I remember worrying about which twin I really was since grown-ups, more often than not, thought I was Alice. If grown-ups were unable to tell us apart, how could I ever hope to know who I was?

I believe that these questions provided a driving force for separation and individuation from my sister, in addition to the normal developmental task of separation and individuation from my parents. I do not know if this situation is true for other sets of twins, or is simply unique to my family system. Alice seemed happy being a twin, so I felt very much alone in my need for individuality. I lived my life in tandem with my twin until Junior High School, where I had the opportunity to attend a few different classes and to make some friends on my own. I loved the freedom.

Being an identical twin continually led me to question my sense of personhood and Selfness. Somehow, I was aware that we were more than mere carbon copies of one another, and I made every effort to communicate this. Alice and I were genetically alike, and yet we expressed different attitudes and values. We looked alike, but we did not think alike. Our uneasy alliance continued until I was married, at age 18. We were no longer "The Twins." It was now "Joyce and Gil"... and, Alice. I felt triumphant! From that moment on, I began to explore my individuality with a sense of freedom previously denied. I passionately enjoyed being Joyce rather than being a twin. I chose to be "one" but not the other. Whenever possible, I denied, ignored and repressed this strong bond and identification to my twin. Twenty years passed before I was able to fully embrace my identity as a twin, and to acknowledge that it was a fundamental aspect of my being. Today, I am able to move in and out of both identifications... my individual self, Joyce, and my twin-self, Joyce-Alice... with ease and comfort. In order to become whole, I needed to first separate, and only later to embrace twin and Self.

In my early thirties, I gave birth to a set of fraternal twins, Bruce and Gabrielle. Now I explored issues of identity, twinship, personhood, and Self from a parent's perspective as well as a twin's. As a Psychologist, I constantly address identity issues in my practice. As a Neuroscientist, I encounter these issues in the debate over nature vs.

nurture. And I discover theological constructs of self, psyche, and soul everytime I explore my own spirituality. Everytime I am in the company of another, I encounter ideas about Self and Being. I wonder how we are alike and interconnected as human beings, and I am interested in how we differ, as well. By understanding and accepting our Self and its multi-faceted nature, we learn to relate in more healthy ways to other Selves. By acknowledging and honoring our differences as well as our similarities, we begin to live in peace.

"ONCE UPON ASOUL" also explores the role of the observer in the creation of its multi-faceted, multi-leveled, and multi-dimensional model of Selfhood. My life, its events and experiences, have profoundly influenced my understanding and perceptions of personhood. Recently, physicists have indicated that the observer is intrinsically linked to what is observed, a phenonmenon that is known as the "Observer Effect." Theology expresses this idea as well, "Quidquid recipitur, ad modum recipientis recipitur," which means, "Whatever is received, is received according to the manner of the receiver." Thus, my ideas and understanding of Self, psyche, and of reality itself, are very much a reflection of the very WHO I happen to be. It is impossible to integrate and synthesize the findings of psychology, science and spirituality without including myself. Essentially, this book and I are inseparable. Nor it is possible for you to understand and relate to this book without also bringing to it the sum of your own life's experiences and events. Rather, author and reader, You and I, are also interrelated and inseparable.

Contents

Acknowledgments	v
Preface	vii
An Invitation	xvii
Menu	xix
A Guide to The One We Are	xxi

Section One
The Evolution of the Psyche — 1

Section Two
The Matter of Science — 7

Old and New Paradigms of Science	9
What Happened Before the "Before"?	10
The Breath of The Invisible	12
Aware of A "Where" Somewhere	13
Non-Locality	15
Non-Ordinary Experience—Alice	16
The Hologram	18
Morphogenetic Fields	21
Synchrony	22
Physics Visits the World of the Shaman	23
Non-Ordinary Experience—Egypt	24
It's About Time	26
Moments in Time	26
Different Ages, Different Stages	28
Communion	29
Separation	31
Classical Science	32
Chaos and Uncertainty	34
Everywhere a Quark, Quark	34
Uncertainty	36
Complexity	37
Union and Reconciliation	40
$E = MC^2$ and the Cosmic Dance	41
The Dance of Shiva	41

Section Three
Slaying the Dragon — 49

- The Gods Return Home — 50
- Ontogeny Recapitulates Phylogeny: The Cycle of Life — 52
- The Evolution of the Psyche — 53
- The Crisis of Oedipus — 54
- The Feminine Principle — 56
- The Process of Transformation — 58
- As Above, So Below — 60
- Different Dimensions, Different Perspectives — 61
  - A Word of Caution — 61

Section Four
The Many Flavors of Psychology — 69

- Once Upon ASELF — 70
- Life With The Goddess — 70
- The Patriarch Emerges — 73
- A Separate Self Emerges—School Years — 74
- Adolescence — 75
- High School — 75
- College — 77
- A Family of Inner Selves — 79
- Healing A Family — 80
- ASELF Meets Science — 81
- In Love — 83
- Wedding Bells — 84
- Chaos and Uncertainty—A Marriage Ends — 84
- The Work of Psychotherapy — 85
- Perennial Philosophy — 85
- The Archetypes — 88
- Stirrings of the Feminine — 91

Section Five
The Realms of Spirit — 97

- The Path of the Heart — 98
- The Yin and Yang of Perception — 99
- Spirit Speaks — 102
  - Story 1 – A Lesson in Receptivity — 103
  - Story Two – The Path of Enlightenment — 104
  - Story Three – The Veil of Forgetting — 105
- Perennial Philosophy — 105
- Many Mansions — 107

The Four Worlds	108
World of Origination	109
World of Creation	109
World of Formation	110
World of Manifestation	111
Awakening Spirit	112
Pathways	113
Holomovement	121
Super Implicate Order	122
Implicate Order	123
Explicate Order	123
A Round Trip	124
Descent into Matter	124
Ascent and Evolution	125
Simultaneous Time and Multiple Worlds	125
Many Lives	128
Karma and Reincarnation	128
Maya—World of Illusion	130
Joe —A Game of Virtual Reality	130
The Art of Dreaming	132
Dream Time	134
Beam Me Down	135
Revelation	135
Gnosis	136

Section Six
The Circular-Spiral Matrix   143

Visions	146
Dream 1—Genesis	148
Dream 2—Evolving Universe	148
Dream 3—Life Begins	149
Dream 4—New Life Forms	150
Dream 5—Evolution Accelerates	151
Dream 6—Physical Birth	153
Dream 7—First People	153
Dream 8—Psychological Birth	154
Dream 9—Exodus	155
Dream 10—Separation	155
Dream 11—Warning Signs	156
Dream 12—Crisis	156
Dream 13—Unfolding	157
Dream 14—Circular-Spiral Matrix	157
Dream 15—Emergent Feminine	159

Dream 16—Sacred Chrysalis	160
Dream 17—Spiritual Birth	161
Dream 18—Sacred Marriage	163
Dream 19—Crossroads	164
Destiny	166

Section Seven
2070 A.D.—The Wisdom Age — 171

Prayer	174
Appendix I	175
The Myth of Demeter and Persephone	175
Significance	176
Appendix II	177
Appendix III	178
Glossary	179
Index	195

Figures and Tables

Figures

1. Multi-Dimensional Selfhood — xxiii
2. Creating a Hologram — 19
   Michael Talbot, "The Holographic Universe," Harper-Collins Publishers, 10 East 53rd Street, New York, NY 10022, p. 15, 1991
3. The Great Chain of Being — 63
   Ken Wilber, Editor, Introduction: Of Shadows and Symbols in "Quantum Questions," Shambhala Publications, Inc., 314 Dartmouth Street, Boston, MA 02116, 1984, pp. 1516
4. Bi-Modal Model of Consciousness — 100
   Arthur Deikman, M.D., presented at Conference, 1994, "Toward a Scientific Basis for Consciousness," April 12–17, 1994, The University of Arizona, Tucson, Arizona, c/o Dr. Deikman, Department of Psychiatry, University of California – San Francisco, 15 Muir Avenue, Mill Valley, CA 94941
5. Jacob's Ladder — 115
   Z'ev ben Shimon Halevi, "Kabbalah: Tradition of hidden knowledge," Thames and Hudson, Inc., 500 Fifth Avenue, New York, NY 10110, 1992, p 41
6. Worlds Within Worlds — 116
   Z'ev ben Shimon Halevi, "Kabbalah: Tradition of hidden knowledge," Thames and Hudson, Inc., 500 Fifth Avenue, New York, NY 10110, 1992, p 41
7. Comparison of Spectrum of Consciousness and Vedanta Psychology (Hindu) — 119
   Ken Wilbur, 1989, "The Spectrum of Consciousness," The Theosophical Publishing House, 306 West Geneva Road, Wheaton, IL 60187
8. Doctrine of Five Sheaths (Koshas) — 120
   Ken Wilbur, 1989, "The Spectrum of Consciousness," The Theosophical Publishing House, 306 West Geneva Road, Wheaton, IL 60187
9. A Game of Virtual Reality — 131
   L. Casey Larijan, The Virtual Reality Primer, Carl Machover (Ed.), 1994, p. 33, McGraw-Hill Inc., 1221 Avenue of the Americas, New York, NY 10020
10. Mandala—Picture 24 — 158
    The collected works of C.G. Jung, Vol. 9. Part I, Bolingen Series XX "Mandala Symbolism" third Ed., 1973 translated by R.F.C. Hull, princton University Press

Tables

1. Three Parallel Life Cycles	58
2. Psycho-Social Development of the Self	58
3. The Great Chain of Being and Worlds Within Worlds	64
4. Perennial Philosophy	86
5. Path of the Heart	98
6. Mystical Four Worlds	108
7. Pathways, Correspondence with Four Worlds	113
8. Four Worlds of Mystic and Scientist	121
9. Evolution's Timeline	152
10. Evolution of Psyche, Earthling and Humankind	165
11. Evolution of Old and New Paradigms	167

Plates

1. Butterfly — 1
   Photograph supplied by JAK
2. Spiral Galaxy — 7
   The Observatories, Pasadena, California
3. Dragon — 49
   Photograph supplied by JAK
4. Encircled Person — 69
   Jose and Miriam Arguelles, "Mandala," 1972, Shambhala Publications, Inc., 314 Dartmouth Street, Boston, MA 02116
5. Yin/Yang — 97
   R. L. Wing, The I CHING Workbook, p. 13., Double Day and Company, Inc., Garden City, NY, 1979

6. Mandala of Vibrations   143
   Heita Copony, "Mystery of Mandalas," 1989, p. 69.,
   Theosophical Publishing House, 306 West Geneva Road,
   Wheaton, IL 60187
7. Oblation   171
   Art: Rob Schouten, Great Paths Publishing, Freeland, WA 98249

Poem   *161*

Andrea B. Cagan, "Caterpillar Dreams," 1994,
Los Angeles, CA

An Invitation

Are you ready for high adventure... through the lands of science, psychology, and the realms of spirit... and ultimately beyond the universe? Would it surprise you to discover that the different languages in each of these lands tell a similar story? Would you like to possess a modern "Rosetta stone," a so-called universal translator to comprehend these common truths and springboard to a life of greater insight, understanding, and enlightenment... universal personhood? Are you really ready for all this? You are! Well then, here we go!

Once Upon ASOUL... long, long ago... there was a great and mysterious... and then there was a wonderful and magical happening... and you and I were part of it all... and so it happened.

Menu

Section One, The Evolution of the Psyche, is an introduction to the ideas that shape and give rise to this book. We learn that the human psyche is governed and directed by different organizing principles throughout humankind's development and evolution.

Section Two, The Matter of Science, considers the instantaneous moment of creation called "The Big Bang," and ponders how the universe exploded out of no where and no when. We learn that our shifting perceptions of space, time, and matter are responsible for the creation of different scientific paradigms and civilizations throughout humanity's history. Four epochs during humankind's sojourn upon the Earth (Communion, Separation, Chaos and Uncertainty, and potential Union and Reconciliation) parallel the development of the human psyche as it progressively evolves through the life cycle of both the individual and the species.

Section Three, Slaying the Dragon, develops the premise that the human psyche is governed and directed by various organizing principles. Shifts in these principles throughout the individual and collective life cycles guide our development from infancy through old age. Essentially, the psychic development of every child recapitulates and repeats the psychological development of the species and the evolution of the psyche. Through the process of transformation, reconciliation of separate and diverse elements within the human psyche occurs. Resolution of duality culminates in a "Sacred Marriage" of masculine and feminine archetypes. Union allows a richer repertoire of possibilities to emerge.

Section Four, The Many Flavors of Psychology, follows a gentle personality, ASELF, through childhood, adolescence, and adulthood. By learning to transcend a number of psychic levels, ASELF uncovers her multi-faceted, multi-leveled personhood to reach the transpersonal. The story of ASELF illustrates the many and arduous psychic tasks that each of us is called upon to master throughout our lives. *Once Upon ASOUL* paints a portrait of ALIFE, none other than a variation of mine and yours.

Section Five, The Realms of Spirit, explores non-physical dimensions of existence which underlie and give rise to our physical world. Encounters with Spirit carry an indelible stamp of certainty and last a lifetime. In mystical and spiritual experiences, subject and object become one. In the act of knowing, duality ceases. The path from

Self to SELF moves from the world of manifestation to the world of formation, then on to the world of creation, and finally opens to the possibility of glimpsing the world of emanation. While we cannot fully know the unknowable, it is possible to gradually approach some of these hidden dimensions, to interact with them, and thereby expand our awareness. When one enters the realm of Consciousness itself, paradoxically, one understands that there are no levels, no dimensions, no higher, and no lower. ALL IS ONE.

The Circular-Spiral Matrix of SELF, Section Six, reveals the unity and connectedness of all existence. Within a circle diversity and unity co-exist. As the seed is part of the tree, we are all part of the whole. The circular-spiral matrix of SELF is the underlying field which generates and sustains all of existence. The center is nameless and eternal, a place of self-renewal and rebirth. The Center is every Self's goal. By reaching into the depths of awareness, we arrive at the underground stream in which all SELVES are united.

In Section Seven, we share a vision of a new era, the world of 2070 A.D.—The Wisdom Age. The people of this era live in balance and harmony with the cycles and rhythms of the cosmos, and have learned to respect all forms of life and Consciousness. Mature and humane solutions to conflict and discord have been discovered and are practiced universally. 2070 A.D. already exists within the realms of possibility; it can be humankind's future destiny, if only we choose to make it so. Dare we seize the opportunity?

There are as many ways to read *Once Upon ASOUL* as there are readers. However, it is suggested that all readers begin with Section One.

Select All: This option follows the step-by-step development of ideas and concepts to reveal the "Big Picture." Proceed linearly and sequentially from Sections One through Seven.

Science: Begin with Section Two – The Matter of Science. Proceed to Sections Six, Three, Four, Five, and Seven.

Psychology: Begin with Section Three – Slaying the Dragon. Proceed to Sections Four, Five, Two, Six, and Seven.

Spirituality: Begin with Section Five – The Realms of Spirit. Proceed to Sections Three, Four, Two, Six, and Seven.

A Guide to The One We Are

Whenever you begin a journey, it is a good idea to take a map or guide along with you. It makes the journey easier and the way less hazardous. A map guides your journey from beginning to end, and enables you to reach your desired destination. This guide to the One We Are introduces the book's essential definitions and characters. It will facilitate your understanding and your journey. For your convenience, these definitions also appear throughout the book and the glossary.

Cast of Characters

ALL THAT IS: ABSOLUTE, ATMAN, CREATOR, DIVINE, GOD, NUMINOUS, SPIRIT, SOURCE.

Ananda: ENOUGH's child, ASELF's step-child.

Annie: ASELF's psychologist.

APSYCHE: Also called the Mind, the totality of all psychic processes, conscious, and unconscious. In this book, it is the part of consciousness emanating from Soul. APSYCHE serves as a messenger between Soul and other levels and dimensions of Selfhood (manifest and unmanifest).

ASELF: Earthly personality and heroine of our story. She is a fictitious character who dwells in the world of space-time and matter. She is a variation of me and you.

ASOUL: ASELF's Soul, personalized for purposes of our story. Determines major life challenges and events that ASELF will experience in the world of physicality. Has endowed ASELF with free will and choice. ASOUL seeks to unite with ASELF so that their goals and needs become one.

Blessing and Harmony: ASELF and Rising Star's children.

Enough: ASELF's second husband, Ananda's father.

Father and Mother: ASELF's parents.

Father Sky: Heavens and surrounding sky and atmosphere.

The Feminine: Archetypal organizing principle which is receptive, intuitive, relational, transpersonal and seeks conciliation and union with the Masculine. The two together confer wholeness and balance to psyche.

The Masculine: Also known as the Patriarch. Archetypal organizing principle which governs the stage of Personal/Ego consciousness and adolescence. When it matures and works in concert with the Feminine, psyche becomes whole.

Matriarch: Also known as Great Goddess and Great Mother. Archetypal organizing principle which governs psyche during pre-personal stages of consciousness and childhood.

Mai: ASELF's yoga and meditation teacher.

Mother Earth: Our planetary home, Nature, our world.

Rising Star: ASELF's first husband, Father of Blessing and Harmony.

The Patriarch: Also known as the Masculine. Archetypal organizing principle which governs psyche during stages of Personal/Ego consciousness and adolescence.

The Sacred Marriage: Developmental stage in which masculine and feminine psychic organizing principles unite and cooperate to create a larger gestalt of awareness and wholeness, bestowing enhanced perception and understanding.

Transpersonal Human: This is the mature, adult psyche governed by both Feminine and Masculine archetypal principles, which endows wholeness and balance. Ego releases its centrality and assumes an ex-centric focus. Transpersonal human is a responsible co-creator with SOUL.

Assumptions

1. All divisions of the SELF are arbitrary, for all are portions of a greater whole.
2. Since SPIRIT pervades and interpenetrates all dimensions of reality and existence, we are spiritual as well as physical and psychological beings. Our task is to acknowledge all these aspects of SELF and to integrate them in order to cooperatively join with SOUL in creating fulfilling, harmonious lives.
3. The three portions of Selfhood—SOUL, PSYCHE, and Self—create a tripartite unity.

A Guide to The One We Are

Figure 1
Multi-Dimensional Selfhood

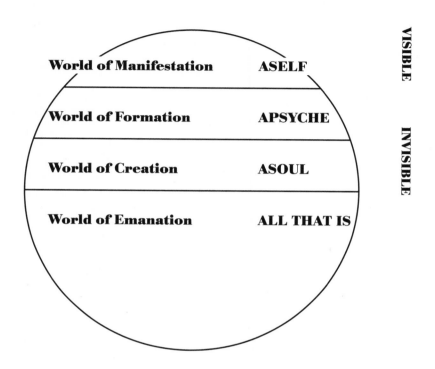

An iceberg provides a useful metaphor for the circle above. Over 80% of this enormous glacier is submerged beneath sea-level. Similarly, only the manifest portion of our multi-faceted, multi-leveled, multi-dimensional SELFHOOD appears in the world of physicality. The rest of our SELF remains invisible and hidden in the non-material realms of Spirit. All portions of SELF are part of the Whole and the One We Are.

The Evolution
of the Psyche

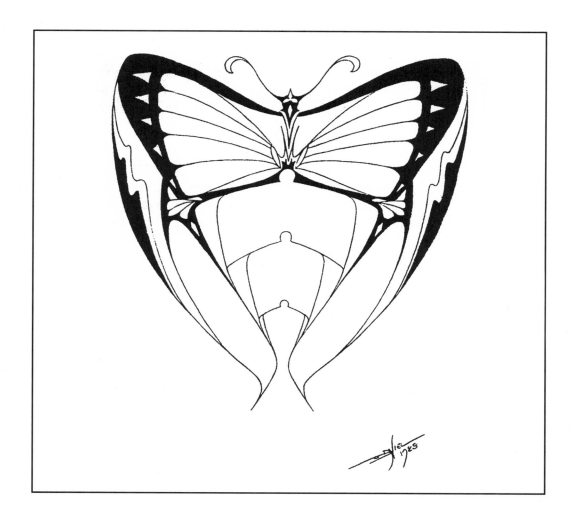

Section one

nce upon ASOUL is a story about a sacred Consciousness which creates space-time and matter and All That Exists. It lovingly bestows its offspring with the opportunity to explore multitudinous levels and worlds of existence. This story follows ASELF's journey into the world of physicality. As her life unfolds, she learns that she is a multi-faceted, multi-leveled, multi-dimensional SELF dwelling within time and space, as well as beyond our earthly boundaries and limitations. ASELF uncovers an enduring, eternal core deep within her Being, and acknowledges ASOUL as her greater SELF. She learns that the human psyche, our entire species, and each individual develops and evolves through certain stages during their earthly sojourn. Each evolutionary stage endows humankind with a particular worldview and understanding of personhood.

As ASELF explores the worlds of science, psychology, and spirituality, she discovers that ideas and concepts in each discipline complement and reaffirm the others. Old and new paradigms about Self and world are converging; gradually a new perception of reality is unfolding. ASELF perceives that humankind is standing at the threshold of a new evolutionary stage of awareness, about to forge a brave new world. *Once Upon ASOUL* invites everyone who recognizes that there is more to heaven and earth than the readily visible and who yearns to explore the deeper mysteries of life, to share the collective wisdom of science, psychology, and spirituality as it integrates mind, body, and spirit into a satisfying wholeness.

We cannot separate our understanding of reality from ourselves, no matter how hard we try. *Once Upon ASOUL* reveals how humanity's present world-view and perceptions of Self and Universe reflect our current stage of evolutionary development, as well as our earlier decision to separate ourselves from the rest of nature. Our outer world faithfully echoes our inner feelings of uncertainty, alienation, and despair. We live in a world that has ignored the sacred and no longer recognizes its connection to SPIRIT. Once upon a time, humankind left the protected and sacred circle of its ancestors to explore a unique and separate consciousness. *Once Upon ASOUL* urges humankind to awaken and to acknowledge its spiritual roots so that we may heal ourselves and our planetary home.

> "We are living in what the Greeks called Kairos—the right time—for a 'metamorphosis of the gods.' ... so much is at stake and so much depends on the psychological constitution of modern man."[1]

Once Upon ASOUL was also born out of a need to express what I have learned as a person, scientist, psychologist, and mystic. It expresses my discovery that all of these seemingly different pathways lead ultimately to common ground. This book provides a synthesis of recent scientific findings which support and enhance our psychological understanding of Selfhood, and integrates ideas found in psychology, science, and the ancient spiritual traditions. By combining the ideas of each of these disciplines, we gain a greater sense of personhood capable of embracing duality, opposites, and paradox. We open to diversity, no longer needing to deal only with one way, or "all or none" thinking. We embrace the sacred and provide dignity to all of humanity and life upon our planet.

Once Upon ASOUL speaks to an enduring, eternal core of Consciousness which each of us carries deep within our being. It proposes a multi-faceted, multi-leveled, and multi-dimensional model of Self. Humanity's rich, multi-dimensional Selfhood is revealed and allowed to unfold. This book suggests that the creation and development of Selves or aspects of personhood is a natural, innate talent of the human psyche which continues throughout one's entire life span. While some aspects of Psyche may be pathological and problematic, others carry positive potential for healing and creativity. Consolidation and integration of ideas taken from science, psychology, and spiritual disciplines help us to achieve synthesis of mind, body, and spirit, and enable us to experience and know a greater wholeness. *Once Upon ASOUL* invites all of us to participate with Consciousness, Itself, as we explore the sacred mysteries beneath existence.

We learn that Psyche, as well as all humanity, experiences different evolutionary stages of development during its sojourn upon Earth. Each stage of psychic development is governed and directed by specific organizing principles. Shifts in these principles throughout individual and collective life cycles guide our development from infancy through old age. Each particular evolutionary stage influences and determines our perception of reality and understanding of personhood. Essentially, the psychic development of every child recapitulates and repeats the psychological development of humankind and its evolving psyche. All stages are interrelated; successful resolution of each preceding stage is required before the next higher level of development can unfold. Reconciliation of separate and diverse elements within Psyche facilitates the unfolding of a richer repertoire of possibilities. Achievement of wholeness through the resolution of duality is called the "Sacred Marriage."

Renowned scholars and scientists have begun to uncover an incredibly strange, Sci-Fi landscape of subatomic particles that include "charms, quarks, and gluons" as well as parallel and multi-dimensional universes. They speak of the duality of particle and wave, the Observer Effect, the Uncertainty Principle, and a reality radically different from the one recognized through our egos and five physical senses. Modern scientific findings are providing remarkable confirmation of ancient spiritual and shamanistic concepts of reality. Collectively, newly emerging views of science offer support for the ancient mystical idea of a "microcosm within the macrocosm," and "as above, so below." Shaman and scientist propose amazingly parallel descriptions of our world and universe. Scientists have begun to attribute a sense of sacredness to their discoveries.

Yet, an existential despair and emptiness pervade our homes and communities, precipitating a sense of longing for connection within our hearts. So many people feel bored or restless and awaken one day to realize that their lives are empty. Somehow, somewhere, we have lost our way. We experience little quality of life or joy, moving aimlessly through our days, sensing that something is missing. Indeed, psychology, having once repressed spirituality, is now reconsidering the wisdom and teachings of ancient traditions so that we may return to deeper sources of being and re-experience our Oneness.

The essential focus of *Once Upon ASOUL* is nothing less than an exploration of the eternal, existential questions: "Who am I?" and "What am I doing here?" and the deeper longings behind these thoughts: "How can I heal?" "How can I become whole again?" Recently, authors have explored inner child work as an avenue to satisfy this yearning, and have provided a bounty of valuable information. Indeed, exploration of who and what I am has uncovered an ever growing catalog of angry, anxious, contemplative, creative, critical, defiant, depressed, enthusiastic, entrepreneurial, fanatic, frightened, hostile, hurting, inadequate, innocent, loving, magical, successful, timid, transcendental, vulnerable, and wise parts of myself. Which are ME? I ask. The answer is: All of them! Inner Selves come in all flavors. Our task is to bring into awareness each of these varied parts, not only the ones that are pathological and destructive, but also those aspects that carry the potential to be positive and healing.

However, inner child work primarily addresses the personal unconscious within a four-dimensional person who lives in a world of space and time. What is additionally needed for healing and growth is to adopt a model of Self and Consciousness that transcends the boundaries of the space-time continuum. Our task is to embrace a SELF that dwells within the invisible, transcendental, and mysterious territories of the psyche. Unless we acknowledge the more spiritual and sacred portions of SELF, we will never truly heal or be "whole."

Until we infuse mystery and awe into our understanding of personhood, we will never realize our fuller potential as human beings. For we have reduced ourselves to mere fragments, caricatures, and slivers of "unwholey" individuals, devoid of meaning and purpose. What is required for wellness and healing is an integrated and larger sense of who and what each of us is. *Once Upon ASOUL* assists us in this endeavor.

Suppose that the "you" that you think you are is really a multi-faceted, multi-leveled, multi-dimensional "being." You might wonder if you can see, hear, feel, and interact with any of these other aspects of being. Are these other portions of the psyche real or imaginary? Are they alive in the same sense that I consider myself alive, or are they instead mental constructs that enable me to understand my sometimes contradictory nature? And what if you become so fragmented that you are like "The Old Woman in the Shoe," with so many parts and so many Selves that you...? Do parts of you, originating in the collective unconscious, belong to me as well? Who in fact are you?

I propose that the human psyche has a natural, innate ability to create multiple Selves throughout one's life span. Each of us has the ability to become conscious co-creators with our psyches in designing lives that support not only the meeting of everyday goals, but also the realization of a vastly greater potential. Conscious participation in the creation and integration of different Selves on multiple levels enables us to live more harmonious, fulfilling, and enriched lives. *Once Upon ASOUL* posits the notion that we can change the quality and direction of our lives by choosing to change ourselves, that each of us can change our past and our present, as well as custom design our future. The ability to continually seed new aspects of personality throughout our lifetimes provides avenues for future growth, development, and personal evolution. The necessary resources are found deep within every one of us. Indeed, they are part of our human heritage!

Once Upon ASOUL develops a circular-spiral matrix as both ground and form for the development and expression of a multi-faceted, multi-leveled, multi-dimensional Self. Within a circle, there is unity; there are no "lower or higher"; all are part of the whole. In the deeper recesses of Consciousness, there are no better or worse parts of the Self, since all stages and all aspects of SELF have validity. The seed and the tree are understood as one and the same, both being aspects of an event occurring over time and space.

By traveling to various territories of the psyche—beyond our ordinary four-dimensional world of time and space—we gain assistance from still other aspects of a collective psyche or consciousness that gives rise to our everyday world. We are encouraged to journey more deeply into our own being and consciousness, so that we may more fully participate in the process of creation Itself. We, as a species, have an opportunity to evolve a Conscious Collective that joins both the personal and collective unconscious in shaping our world.

The interplay of diversity within unity, and the many expressions of ONE are the challenge awaiting humanity. I invite you to join me in developing portions of Psyche that will heal not only ourselves, but our lovers, our families, our planet, and our universe, as well. In the process of recognizing and reconciling our opposites, we begin to heal ourselves. As we cultivate and explore our uniqueness and allow expression of our individual differences, we become less threatened by the differences of other cultures, religions, and races. As each individual grows in personal understanding and responsibility, one realizes there is no need to blame, to fault, to judge, to plunder, or to wage war. Any "wars" or conflicts will be resolved within our individual psyches. As we take responsibility for our own views and actions, there is no further need to project our individual and collective fears, jealousies, hates, or lusts upon any "other." We extend our capacity to be truly human when we cooperate rather than compete, and express rather than repress our many and varied voices.

An evolution of Self and humanity will be reflected out into the world, creating new forms in education, family, government, medicine, psychology, religion, science, and the arts. Through the development of a deeper awareness of Self, we can choose and we can heal. Each of us can more fully realize the ONE WE ARE!

Notes

1. C. G. Jung, *The Undiscovered Self* (New York, The New American Library, 1959), p. 123.

The Matter of Science

Section two

erhaps you are puzzled about the need to explore old and new paradigms of science in a book about the SELF. Science has already taught us about the bones and muscles that move our bodies through space and time, how we genetically reproduce our young, and the biochemistry of metabolism. Science has enabled us to send space probes to the moon and other heavenly bodies, encouraging our vision to go "where no one has gone before." But what can science possibly teach us about a multi-faceted, multi-leveled, multi-dimensional SELF?

Our views of science reveal our present understanding and conception of ourselves as we dwell within a world of space, time, and matter. Humankind has projected a certain understanding and view of Self upon its sciences, that, in turn, is faithfully reflected back to us. Our science reveals who we believe ourselves to be.

Many of us are still looking to change the world out there. We might be able to improve our air quality and reduce the pollution of our lakes, rivers, and streams; we might slow the advancing desert and stem soil erosion, and we might succeed in recycling our resources if we concentrate our efforts outwardly. All of this will prove beneficial and is, indeed, necessary. Yet it will not sufficiently change our world or our understanding of science in time to avert disaster. Rather, the changes we must make—and soon—must occur within. We must first change our views of ourselves, our understanding of nature, as well as our purpose and our place within the cosmos before we can hope to significantly alter the world in which we live. We must rebuild our beliefs of Self to embrace a more wholistic and harmonious understanding of ourselves as part of the fabric of existence. We can no longer pollute our world without serious consequences. We can no longer exploit earth's resources and bounty for our own ends. We can no longer demean people of other races and cultures. And we can no longer live apart from nature, nor continue to ignore its limits and violate its laws. This is the lesson that science is beginning to learn and to reflect back to us.

Old and New Paradigms of Science

The evolution of the human psyche and humankind's understanding of its world evolve together. A look at old and new paradigms of science will help us recognize their impact upon our present world-views, and how they shape humanity's perception of the universe in which we live. Until the 1600s, the Ptolemeic "geocentric" view of the universe prevailed; both humankind and earth were considered the center of the universe. But Copernicus, a Polish astronomer, deftly proved that earth was, instead, moving around the sun (heliocentric view). The Copernican Revolution gave birth to classical, old paradigm science, which has ruled the scientific community until the present. Copernicus's observations and elegant mathematics paved the way for Newton's discovery of universal gravitation. The work of Kepler, Bacon, Darwin, Galileo, Newton, and Descartes collectively created the prevailing view of "old" paradigm science. Classical science envisions a mechanistic, predictable, orderly cosmos, and a material world of discrete, separate objects connected by a sequence of causal events. In theory, all can be predicted and controlled. It is a deterministic science.

"New" paradigm science perceives and recognizes a different universe. It is holistic, suggesting that all existence emerges from a non-material field of energy which supports and sustains Universe. Hence everything is interconnected and interrelated, rather than discrete and separate. Earth is considered a living, aware organism; it cannot survive in fragments. New paradigm science concerns itself with wholes, systems, newly-emerging properties, chaos, duality, uncertainty, indeterminism, and paradox. In "new" paradigm science, everything is considered in relationship to everything else, and the whole is greater than the sum of its parts. New paradigm science emerges out of the discoveries of Bohr, Pauli, Einstein, Schrodinger, Bohm, and the fields of quantum physics, complexity, chaos, hyperspace, and modern cosmology.

These two scientific views seem to be polar opposites of one another; in truth, they are complementary. Each has achieved significant success in revealing the ways of nature in our world. Nevertheless, a new understanding of humanity and universe is emerging; a radically different scientific view of wholes, relationships, systems, and connections is unfolding. A more holistic world-view will be of tremendous assistance in humankind's efforts to heal Self, family, nation, and planet.

If we wish to inhabit a world that can support and sustain life, it is humankind that must change. Any changes that we make will be projected outwardly to leave their imprint and consequences upon the global canvas. Whatever changes we make will be accurately incorporated into our feedback loop with nature, and will most certainly influence and affect the entire web of life in which humanity resides. There is no other way. To change the world, we must first change ourselves.

Let's begin at the beginning, with science's description of an extraordinary event which occurred prior to existence and *before the before* of any space or time, an event that led to the origin of an entire universe. Surprisingly, science's most recent explanation, the "Big Bang," bears a striking resemblance to the creation myths of earlier civilizations and spiritual traditions. Old paradigm science followed the linear arrow of time forward, from the origin of our universe into humanity's present. It concerned itself primarily with the laws of nature and the physical world of matter.

What Happened Before the "Before"?

Once upon a time, there were no seconds, minutes, or hours; no days, weeks, months, or years; there was neither here nor there, no earth, no moon, no sun and no universe. There was, instead, an eternal, timeless void. Suddenly, we are told, there was an explosion known as the "Big Bang," and in an instant time, space, matter, and all that we know as reality—began. And in a flash, a universe was born from an infinitely hot and dense point of zero-matter known as a *singularity*.

Such is the creation story presented by modern cosmologists and scientists to explain the unexplainable origin of our universe; a reality that seemingly was created out of nothing. Now let us turn to some other creation stories proposed by a number of spiritual traditions throughout humankind's history.

From the Babylonian *Enuma Elish*:

> "When a sky above had not been mentioned
> (And) the name of firm ground below had not (yet even) been thought of;
> (When) only primeval Apsu, their begetter,
> And Mummu and Ti'amat—she who gave birth to them all—
> Were mingling their waters in one;
> When no bog had formed (and) no island could be found;
> When no god whosoever had appeared,
> Had been named, had been determined as to (his) lot,
> Then were the gods formed within them."[1]

In The Book of Genesis in the Old Testament, we read:

> *"In the beginning God created the heaven and earth. And the earth was without form, and void; and the darkness was upon the face of the deep. And the spirit of God moved upon the face of the waters. And God said, Let there be light; and there was light."*[2]

The Holy Qur'an similarly states:

> *"God is the Light of the heaven and the earth,"* and *"It is God who has created the heavens and the earth, and all between them, in six days."*[3]

The Hopi Indians describe the creation of four consecutive worlds. The first arises out of endless space where there is neither beginning nor end, no time, no space, no life... just an immeasureable void. The Hopi suggest that the beginning and end existed first within the mind of the creator. They speak of solid substance appearing from endless space into forms.[4]

Australian Aborigines describe a brief, eternal primordial event in which the world took form from a formless void, from a timeless reality that existed prior to creation, and still exists in the present. They call this spiritual realm Dream-Time. The Aborigines envision life as a web of interacting particles in which humankind and nature are co-equal partners. Since Dream-Time is eternal, it can be entered and known in the present, something that the Aborigines do during an altered state of consciousness known as "Dreaming" (*Tjukuba*).[5]

Buddhist cosmology is unique in that there is no conception of a Deity or specific creation event. Instead, Buddhists believe in the existence of an eternal "Void." Unconscious karmic activity and forces interact with minute elements or particles within this void, generating friction, fire, energy, and heat, and permitting a chain of life to evolve. The void is a timeless spiritual realm without beginning or end and without form. In this realm there is no time, no past, and no future, only an enduring, eternal present. In the *Tao Te Ching* we read:

> *"In the beginning was the Tao. All things issue from it; All things return to it."*[6]

Despite the varied and numerous creation stories, there is remarkable agreement between theology and science that somehow our universe emerged out of no where and from no thing, giving birth and form to all of reality. Scientists have yet to determine who or what initiated the Big Bang. Instead, they concentrate upon the events immediately following this explosion and upon the way Universe works. Their cosmology, like Universe, begins only at the moment of the Big Bang. In contrast, theologians explore the "how" and "why" Universe began many eons ago. Every spiritual tradition acknowledges and recognizes a Source or a Creator and refers to this being as "God."

The Breath of The Invisible

Perhaps, like most of us, you have never seriously concerned yourself with thoughts of space or even considered what space is, or what it might be moving into as the universe, itself, expands. Most of us merely live and move through space without giving any special thought to its purpose or existence. Yet, space is everywhere. It permeates and percolates through our world, completely surrounding us, while seemingly separating "us" from any "other." Mystics tell us we are actually embedded and immersed within space, as well as intrinsically joined, one to the other. We usually think of space as being empty, but it really isn't. Rather, it is filled with myriad infinitely tiny particles, too small for the human eye to perceive. Space contains viruses, bacteria, dust, photons of light, gravitons of gravity, subatomic particles, radio-waves, and cosmic dark matter, to name but a few. Space also contains virtual particles and potential matter, that under special circumstances and conditions become manifest for the briefest of moments. Sometimes, it is possible to create "something from nothing." Indeed, space is not empty.

Space is simply there; it provides a place for things to happen. Space also supports the cosmic web of existence, in which all is interrelated and interconnected. Space creates boundaries and relationships, and is part of both inner and outer worlds, helping to form and define our subjective and objective realities. We think of physical space as existing out there, beyond ourselves, while psychological and spiritual space is considered to reside within. Space, as well as time, determines the rhythms and cycles of our world, our speech, our music, our art, and our literature. Space enables us to comprehend the spoken word, to read and understand the letters upon a printed page, to transform sound into music, and to define and make sense of our world. Without space there is no place for a universe, and no way to reach out and touch one another. Indeed, space provides the very breath of life and supports the ground of our being.

Scientists and mystics alike are deeply interested in the concept of space. Recent scientific concepts and notions such as non-locality, the hologram, Sheldrake's M-fields, principles of quantum mechanics, and Bohm's ideas of an unmanifest Implicate Order beneath our physical world, are all in remarkable agreement with experiences of emptiness, nothingness, and ideas of the void disclosed in the ancient teachings of shamans and mystics. Both science and the spiritual traditions reveal deeply hidden, invisible dimensions and inner spaces in which all being and consciousness are seen as interrelated and interconnected. What happens in one realm clearly affects and influences what occurs in any other. Amazingly, more and more branches of science propose ideas and notions about space which support and confirm the teachings of ancient religions and spiritual traditions as well as psychology. Although the approach, techniques, and language of each discipline differ, there appears to be a surprising similarity in their understanding and description of what we call space.

Aware of A "Where" Somewhere

In order to get anywhere in our four-dimensional world (height, width, depth, and time), one needs to know where he is at a particular moment, and where he wishes to be in the next. Hence, location and place are determined by the relative placement of at least two points in space. Without a here, we cannot go there. Without a there, we will never arrive here. Further, the neurological system of an observer interacts with externally perceived objects in order to create an inner experience of places and objects existing in four-dimensional space-time. It is our neurological apparatus which allows us to perceive imperceptibly discrete, small movements occurring in time and space, as continuous and smooth. It is our brain, in here, which interprets sensory information from out there, and determines how we understand and see our physical world.

Some scientists assert that, "The Big Bang was the abrupt creation of the Universe from literally nothing: no space, no time, no matter."[7] Some also imply, "The Big Bang occurred everywhere and was the explosive appearance of space."[8]

Space appears simultaneously everywhere. Most of us would consider space as a substance that differs from ourselves, while most spiritual disciplines, and more recently, a number of physicists, have suggested that we and space are really inseparable. Physicist David Bohm considers space to be the ground of existence that weaves and integrates us into the fabric of life itself. Everyone and everything is embedded within this ground and tapestry of life. Scientists report that our universe has been expanding and inflating ever since the Big Bang occurred. But nobody knows exactly what form of absolute or superspace our universe may be expanding or moving into, or even if a superspace (hyperspace) exists at all. Scientists also want to know whether we live in a closed or open universe, wondering if there is any boundary or an end to our universe. The answer to this puzzle would help them determine whether space is infinite or finite. Thousands of years ago, the Greek philospher Plato and his colleagues considered the same question. To decide, they imaged a hand stretched out to the very edge of the universe, in an effort to ascertain whether there was a wall or barrier at the "end of space." Following this thought experiment, they reached the conclusion that our universe was spatially infinite. More recently, Stephen Hawking, world-renowned astronomer, stated, "If the Universe is really completely self-contained, having no boundary or edge, it would have no beginning nor end; it would simply be" (Hartle-Hawking Theory).[9]

Hartle and Hawking are essentially saying that the universe exists, necessarily! They imply that the creation and validity of the cosmos somehow belongs to realms beyond our limited concepts of space, time, and the Big Bang. Similarly, physicists believe that superspace was never "created," but also exists necessarily, in the same context that spiritual traditions assert the existence of God. For the mystic, space is the "Void"; it is the timeless nothingness—eternally existing, knowable yet indescribable. Ultimately, both theology and science require an infinite, primordial realm that existed, before the before and always, and provides the medium or domain in which things "happen." These ideas are beautifully expressed in the Chinese *Tao Te Ching*.

> *"There was something formless and perfect*
> *before the universe was born.*
> *It is serene. Empty.*
> *Solitary. Unchanging.*
> *Infinite. Eternally present.*
> *It is the mother of the universe.*
> *For lack of a better name,*
> *I call it Tao.*
>
> *It flows through all things,*
> *inside and outside, and returns*
> *to the origin of all things."*[10]

Non-Locality

Non-locality is based upon the notion that all existence is part of a seamless, undivided whole. Existence of a unitary field, which forms and sustains our physical world, implies that our cosmos is not a closed, isolated system. Indeed, Bell's Theorem[11] contends that without the existence of a supporting, non-local realm, there can be no space, no time, and no matter. Scientists imply this invisible, non-material realm *exists necessarily*.

The notion of "non-locality" suggests that, at the sub-quantum level, time and space no longer matter; essentially everything is known at once. Because everything is enfolded within this undivided whole, there can be no separate elements or objects. Indeed, every point is equal to every other point. Scientists reveal that communication is no longer restricted by mass, time, or the speed of light. Rather, information flows simultaneously to everywhere and everywhen, and objects appear to instantaneously leap from one point to another, without the necessity of passing through space or time. Reports of synchronous, faster-than-the-speed-of-light communication between pairs of electrons spatially separated over millions of miles defies the underlying assumptions of Einstein's Theory of Relativity. Einstein's theory forbids anything with mass from moving faster than light in our universe. Scientists tell us that an electron seemingly "knows" instantaneously what its partner is experiencing on the other side of the cosmos. How can one particle of matter, so distant, simultaneously influence another? The difficulty lies in seeking an answer within the realm and laws of physicality. It seems more likely that the paired electrons communicate in or through another order of existence, in which different rules apply, and in which they are seen as inseparable elements of a larger whole.

The search for another dimension, or order of existence, deeply concerned physicist David Bohm. (I emphasize his ideas in this section because they bear a striking resemblance to ideas of the ancient mystical traditions.) Bohm's search eventually led to his proposal of a fundamental, invisible level or field beneath Universe. This unitary field, called the Implicate (enfolded) Order, is both creator and created; it contains and is All. Bohm's Implicate Order encompasses an infinite spectrum of invisible, non-material levels enfolded within levels; it is Source and originator of all objects, events, and experiences encountered in the physical or Explicate Order. Essentially, all physical forms and structures ensue from the endless interplay between these two levels of reality (implicate and explicate, unmanifest and manifest). Bohm, like many spiritual teachers, suggests matter is illusory. He asserts that our neurological systems create the illusion that objects are "real" and separate, and he contends that our brains interpret "our world," endowing it with substance, shape, and form. The idea of an illusory world corresponds to the Hindu and Buddhist concepts of "Maya" which also imply that "We" and Universe are not really "real."

Yet mystics and scientists advise that it is possible to perceive and know the world quite differently, through the exploration of different orders and levels of existence. The mystic alters his state of consciousness (through meditation, chanting, fasting, beliefs), and in so doing, engages a far different reality, while the physicist uncovers stranger and more bizarre characteristics and properties the more deeply he delves into the subatomic world.

Bohm's concepts rest upon the property of non-locality. Essentially, he sees no reason to divide matter into arbitrary, separate elements. Bohm asserts that, at deeper levels of reality, matter eventually dissolves into an immense dimension of formless energy. The existence of an underlying "quantum potential" or field which permeates all of space ensures that everything is interconnected and interrelated to everything else, much like our hand and leg are integral parts of our (whole) body; consequently, everything is a part of the whole. Bohm further contends that this unbroken field coordinates and organizes all the experiences and events of our physical world: "The behavior of the parts was actually organized by the whole."[12] Bohm imbues these deeper levels of order with mystical qualities as he conjures up descriptions of objects containing or blending into one another. Accordingly, the Explicate Order dynamically blends back into the Implicate and out again, to endlessly form our reality, moment by moment.

Bohm also contends "wholeness" is an essential governing principle of Universe. An increasing number of scientists are beginning to acknowledge deeply hidden, invisible orders of existence, in which all of us—indeed, everything that exists—are intrinsically connected and joined. Eventually, it is to these endless hidden, enfolded levels that we must look for the understanding of otherwise strange, non-ordinary, and presently inexplicable events.

Non-Ordinary Experience—Alice

Indeed, the notion of non-locality might partially explain a paranormal event that occurred several years ago. In 1989, my twin sister Alice arrived at my home with lip dangling and leg dragging. It was shocking to see her this way, and we knew that something was terribly wrong. Within hours she was admitted to the hospital and diagnosed with a large, inoperable brain tumor. I felt as if I were in a dream and would, hopefully, awaken shortly. Of course it all proved to be terribly real!

We chose to do a surgical biopsy, in order to determine the type of tumor and if treatment was possible. Alice slipped into a coma and remained in this state for the next four days. During this time, I began to experience an intense despair and anguish unlike any I had previously known. I was surprised and alarmed by the enormity of these feelings and grew increasingly concerned. Finally, I called a friend and asked for help. I explained what was happening and that I didn't "feel like myself." With this thought, the heaviness began to roll and move away. I suddenly understood what was happening. Even though I was at home in Chatsworth and Alice was in a Los Angeles hospital, I had telepathically picked up Alice's despair and anguish while she was in a coma.

Alice and I were in communication with one another, despite a distance of over 50 miles. Once I recognized what was transpiring, I was able to move in and out of these feelings at will. I felt a sense of relief to be again myself. I learned how to move among the sense of Joyce's Self, and to recognize a sense of Alice's Self, and to know each as different states or aspects of me. Although I was aware of the deep agonies Alice was feeling, I was now able to separate her feelings from my own. Miraculously, Alice responded to radiation therapy, even though we were told she would not. She has lived beyond the days, months, and years of her prognosis without any return of cancer.

For me, the meaning of this experience speaks of the deep psychological and biological connections that I share with my identical twin sister. It has long been recognized that incidents of precognition and telepathy occur with great frequency among twins. Throughout our childhood, Alice and I had numerous telepathic experiences. Non-ordinary experiences between twins lend support to the mystic's contention that all of us are interconnected and part of the web of life. Each of us has an innate ability to tune in to the Implicate level and to become more sensitive to the happenings of another. Remarkably, this event found me in one city, and my twin in another, yet we were undeniably, and instantaneously connected. On some level, Alice was aware of her condition and able to communicate her distress and anguish, even though in a deep coma. She wanted someone to know her feelings, her fears, and her pain. Psychic communication was the only form of expression available to her. The intensity of her emotions released a psychic S.O.S. through the universe, seemingly addressed to me. I wondered if she knew she had sent it, and I wondered if she knew that I had received her message. There was little to do, but return to her bedside, pray, and lend support.

I believe that my meeting with Alice occurred within an invisible, non-local realm beyond our physical world. Astonishingly, there are portions of SELF, existing outside of space-time—aware, knowing, and capable of expressing needs in ways not yet understood.

The Hologram

Further support for non-locality and wholeness comes from another notion suggesting that the universe operates according to holographic principles. A hologram, or three-dimensional image, is recorded when we split a single, coherent source of light (i.e. laser) into two separate beams, reflect the first beam off the object, and record the resulting wavefront interference pattern upon a photographic emulsion. A three-dimensional, holographic image of the object is revealed when a second laser beam is passed through this piece of film. Figure 2 illustrates how a hologram is formed.

It is possible to walk around a hologram and to view it from many angles, as if it were a real object. However, if you were to reach out to touch the object, you would find that there is nothing there. What is perceived is merely an *illusion* or image of a three-dimensional object projected into space. Another unique property of a hologram is that even if we cut the film into smaller and smaller fragments, each piece of holographic film still contains information from every point of the object, so that the whole image can be reproduced from any fragment of the recording. This is reminiscent of the DNA molecule, which embodies the entire genetic code of a species, and is found within the nucleus of each and every cell in our body.[*] Each cell thus carries a complete set of genetic instructions that could, in theory, serve as a template to reproduce and replicate an entire being. Both DNA and holograms embody the principle of non-locality; all points are equal and all portions contain the whole.

It is also possible to construct a multi-faceted hologram containing images of many objects, thereby increasing the amount of potential information stored on the film. Neuroscientist Carl Pribram has proposed the intriguing concept that human memory is holographically stored, and that the brain is a hologram. Such a model leads to a profoundly different conceptualization of Self and Universe, and implies that our perception of "objective reality" requires rethinking and restructuring. Perchance, each of us is actually a composite, multi-faceted, multi-dimensional hologram projected or seeded into time and space. Yet, we are not sure what hologram we represent. This

---

[*]Red blood cells do not have a nucleus, hence they do not contain DNA.

Figure 2
Creating a Hologram

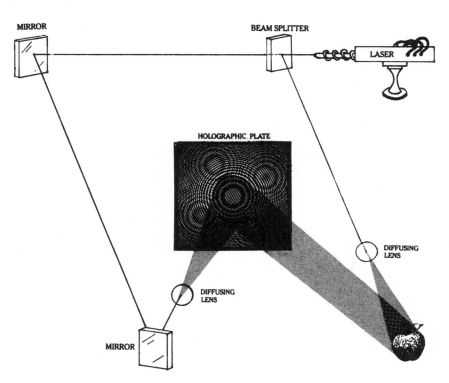

A hologram is produced when a single laser light is split into two separate beams. The first beam is bounced off the object to be photographed, in this case an apple. Then the second beam is allowed to collide with the reflected light of the first, and the resulting interference pattern is recorded on film.

Source: Talbot, p. 15.

uncertainty is one of the greatest mysteries of life. With great daring, other scientists, including Bohm and Pribram, assert that the universe is also a hologram. Because constant motion and change, as well as holographic properties, are at work in the universe, Bohm prefers the term "a coordinated holomovement" to describe the nature and dance of the cosmos.

To the mystic, space is the necessary ground for all existence and experience within the physical world, and the embodiment of mind and consciousness. *Space is the stuff of spirit!* Mystics tell us that "outer" physical space is a reflection of "inner" psychic realms. Hence, one's perception of space and time reveals an individual's state of consciousness and awareness. For example, *pre-temporal* consciousness would involve no awareness of space and time at all. While we are gestating in our mother's womb (peri-natal development), we have little or no conception of earthly space-time or life as we will later experience it. Life simply begins at birth, reminiscent of science's claim that Universe emerged out of the Big Bang. The Chinese consider a baby to be nine months old at birth, since they believe that life begins at the moment of conception. *Trans-temporal* consciousness, however, explores realms beyond time and space while one is simultaneously immersed and anchored within it (the body is "here" and the mind is "there"). This is the teaching of shamans, psychics, mystics, and seers, and others who journey upon the various spiritual pathways that enable one to transcend our four-dimensional space-time reality.

Throughout history, we find references strikingly similar to notions of non-locality, holographic universe (holomovement), and Bohm's conception of a unified quantum potential or field, described above. Many millennia ago, Empedocles, a Greek philosopher, wrote: "God is a circle whose center is everywhere and its circumference nowhere."[13]

The Sufis offer the idea that, "the macrocosm is the microcosm," an idea also found in the Hindu, *The Upanishads*.[14]

In the East, the Hindu *Visvasara Tantra* states, "what is here is elsewhere,"[15] while the *Avatamsaka Sutra* contends that "each object in the world is not merely itself, but involves every other object, and in fact, is everything else."[16] Likewise, Fa-Tsuang, a Buddhist, encapsulates the principles of the hologram in his remark, "The whole cosmos was implicit in each of its parts."[17]

In the West, Chief Seattle (1854) echoes Bohm's ideas of interconnectivity and non-locality: "This we know. The earth does not belong to man; man belongs to the earth. This we know. All things are connected. Whatever befalls the earth, befalls the sons of the earth. Man did not weave the web of life. He is merely a strand in it. Whatever he does to the web, he does to himself."[18]

More recently, Ilya Prigogine received a Nobel prize in chemistry for his discovery of dissipative structures which seem to manifest from deeper, invisible levels of reality which resemble Bohm's Implicate and Explicate Orders. Dissipative structures are open, self-regulating systems that create order out of chaos. Dissipative structures frequently reorganize themselves on higher levels of complexity, and develop and unfold newly emergent properties not present in the original system. Prigogine's work has led to the formulation of Chaos Theory in non-linear systems, discussed on p. 34.

Morphogenetic Fields

Sheldrake describes a hierarchy of nested, non-material fields and patterns which are also found in an invisible realm beneath the world of matter. Morphogenic (M-) fields coalesce around a germ or seed to form the blueprints and patterns which arrange matter into protons, electrons, cells, flowers, birds, and human beings. According to Sheldrake's theory, the manifestation of everything that exists in the physical world is governed and organized by M-fields.

Sheldrake also tells us that the more often a pattern is used and/or repeated, the easier it is to manifest. Reinforced patterns generate "least action pathways."* Such habitually favored pathways make it easier to repeat already established patterns and behaviors. Sheldrake reassures us that with enough energy and intent, a new pathway can be selected and developed. Each time this second pathway is selected, it too is reinforced, making it more likely that it will be chosen again. It seems that once a new form becomes established, it readily spreads and infects other systems, shifting selectivity and momentum in its favor.

Least action pathways provide a possible explanation for apparently synchronous, spontaneous changes, insights, and sudden shifts of consciousness. For example, four different laboratories, each having spent many years investigating a particular problem, all suddenly discover the same solution within three days of each other. Another example tells us that initially, only a handful of brilliant physicists were able to understand Einstein's Theory of Relativity. Yet today's engineering freshmen learn and apply Einstein's theories with comparative ease. Does this imply that today's college freshmen are brighter than Einstein's colleagues, or does this illustrate Sheldrake's idea of least action pathways?

Recently, two independent laboratories reported a revised estimate of the universe's age.[19] The Indiana University group suggested that it is approximately 7 billion years old, whereas the group from Harvard University came up with an estimate of between 9 and 14 billion years. Both universities reported their complementary, parallel findings within days of one another. Both groups also raise a puzzling question. They note that the stars and galaxies are believed to be 16 billion years old. Hence, we are left with a universe that is far younger than its contents. It will be interesting to see how this paradox is resolved.

---

*Least action pathways are nature's path of least resistance. Through repetition and positive reinforcement, it becomes easier and more efficient for energy and experience to travel this path. Least action pathways become habits of nature. Sheldrake reassures us that with enough motivation and intensity, we can form new least action pathways and create our world in quite different ways.

Sheldrake believes Morphogenetic fields embody the collective wisdom and knowledge of earth's countless species. In this regard, M-fields may be considered *biological* counterparts of Jung's *psychological* archetypes. Both M-fields and archetypes are invisible, non-material forms which organize and guide the events and happenings of our physical world. The correspondence between the ideas of these two men is striking. Jung tells us: "There are as many archetypes as there are typical situations in life. The endless repetition has engraved these experiences into our psychic constitution, not in the forms of images, filled with content, but at first only as forms without content representing merely the possibility of a certain type of perception or action."[20]

Synchrony

Jung's concept of synchrony in which a cluster of interconnected events, although not causally linked, share a relationship to one another through archetypal, unconscious psychic processes, lends further support for the existence of a psychological unified field in which all things are interconnected and non-local. Hence, both science and psychology posit the existence of an underlying, non-material unity that gives rise to the events and happenings in our everyday world. Such a synchronous series of events may reflect the efforts of a determined psyche to simultaneously seed three or four *co-incidents* into an individual's world of space-time, in order to dramatically gain Ego Self's attention and bring about desired change.

An example of Jung's concept of synchrony begins with an individual's dream of a scarab (beetle). The next day, as this woman tells her analyst about her dream, a beetle flies into the room. A day later, the woman sees still another beetle in her friend's home. It is noteworthy that in the particular town in which this person lives, it is rare to see even one beetle. These three events are interconnected and interrelated. Synchronous events impart great meaning to those who encounter them, and invite humanity to reconsider its present understanding of time, and how the everyday experiences of our world are made manifest.

In another example, you realize that you haven't heard or spoken to a particular friend for several months. Two days later, you hear his name casually mentioned in a conversation. When you return home, you find a letter from him in your mailbox. The date of your friend's letter indicates that he wrote it on the same day you first thought about him. You sense that in some undisclosed manner, all these events are connected.

Physicist Paul Davies's statement captures the non-local nature of synchronous events: "... non-local quantum effects are indeed a form of synchronicity in the sense they establish a connection—more precisely, a correlation between events for which any form of causal linkage is forbidden."[21]

Physics Visits the World of the Shaman

Physicist Fred Wolf personally applied his knowledge of theoretical physics to the Imaginal world of the shaman. He asserts that through our "intent," we create new "least action pathways." Wolf audaciously implores each of us to use our intent to more consciously and responsibly co-create the experiences of earthly life. According to Wolf, Consciousness unites and sustains all orders and dimensions of reality. Wolf discloses that shamanic and quantum worlds are essentially one; he implies that material and transcendent worlds have a certain shared order. Shamans regularly visit the Imaginal realm in their quest for precious healing and wisdom. Wolf also implies that our greater SELF or "I" is a field of Consciousness, fully awake, aware of Itself, and immersed within all of space-time. "I" is creator and created, the One and the Many.

It is meaningful to consider each person as an event or process occurring over a particular space-time coordinate, rather than as a *discrete,* individual being. Indeed, physicists have posited the existence of invisible fields, forms, and patterns that lie beneath the physical world and organize matter into systems and wholes. It seems that without these forms and templates, there would be no physical world, no you, and no me.

In summation, neuroscientist Carl Pribram suggests, "...we must conclude either that our science is a huge mirage, a construct of the emergence of our convoluted brains, or that indeed, as proclaimed by all great religious convictions, a unity characterizes this emergence and the basic order of the universe."[22]

The concept of non-locality and the idea of an underlying ground beneath all reality may partially explain non-ordinary states of consciousness.* Such altered states of awareness have been reported by people in all walks of life; they often lead to profound transformation and personal growth. Paranormal experiences have also provided hints and glimpses into dimensions beyond our own. Investigations of non-ordinary experiences lead to the development of a radically different

---

*The National Academy of Sciences considers investigation of paranormal phenomena and non-ordinary states of consciousness a valid scientific discipline.

science, as well as to new and expanded understanding of Self and Universe. A non-ordinary experience imbues each person with incredible certainty that a portion of his or her Consciousness eternally exists in other realms and dimensions of being.

Non-Ordinary Experience – Egypt

I, too, have known such experiences, and would like to share an extraordinary event that occurred some years ago, while my husband and I were in Cairo. The Egyptian Minister of Antiquities had graciously invited us to visit a newly uncovered tomb on the south side of the Great Pyramid. This spectacular discovery was so recent that few outside of Egypt knew about it. We were delighted with the opportunity to watch an archeological dig in process. A number of artifacts were brought forth from the tomb for us to view. One object was a remarkably preserved skull. As a neuroscientist, this was hardly the first skull I had ever encountered. Yet, I had a sudden, strange, and compelling desire to hold this skull between the palms of my hands and to recite the Jewish prayer for the dead. Never, in all my years of scientific training and teaching, had I experienced a similar wish to recite a prayer, nor have I felt the need since. At this particular moment the urge was overwhelming and it seemed to come from out of no where and no when.

I do not know if my choice of prayer meant that the person had once been Jewish, or that this was simply the only prayer of the dead that I happened to know. The wish to recite a prayer as well as to hold the skull became more intense, so much so, that I decided to share my desire with the others accompanying me. They grew upset, and reminded me that I was a stranger in this land, and that my ways differed from those of my Egyptian hosts. Even my Egyptian guide asked me not to approach the skull. Respectfully, I refrained from doing so, yet the urge to recite the prayer remained with me. Six hours later, I finally decided to honor this need, and went into a quiet, dark room. I closed my eyes and silently recited the prayer for the dead, while I visualized myself holding the skull between my hands. As I finished, a shudder went through my body, followed by a great sense of calm and serenity. I felt as if I had completed something important and sacred.

I have often pondered the meaning of this experience. It seems related to the ideas of interconnectedness and interrelatedness, as well as to the holographic property of memory and the notion of non-locality. I realized that it is possible for one to participate in events that seemingly occurred long ago, implying that all time is simultaneous rather than consecutive. The skull and I were connected in ways that had nothing to do with space and time as we understand it. Somehow both of us were needed to complete this event and it didn't matter how many millennia intervened. Perhaps the one who lived and died thousand of years before understood that one day another individual, from another age and another world, would come along to affirm her death. Egypt was our meeting point.

Noticeably, no one else at the site felt a similar urge or association with this skull. I was the only one who felt compelled to act as I did. I've wondered if this experience was specifically intended just for me, or whether anyone else who responded to the call would be equally acceptable. I may never know. Nevertheless, I often feel that I received an invitation to play a special role in a cosmic drama that awaited and anticipated my arrival in Egypt many thousands of years after the event occurred. I felt honored, awed, and blessed.

Ever since the occurrence of this paranormal event, I have found myself drawn to issues of death and dying, often working with individuals who are experiencing life-threatening or terminal illnesses. Whenever one engages their illness and/or dying, they experience remarkable growth, completion, and great serenity. So I feel it is a privilege to assist and guide people through this intense, often final stage of life. My decision to enter and to work with the ill and dying seems to be a gift and a legacy from my mysterious encounter in Egypt.

My own encounter with the paranormal, as well as similar experiences disclosed by so many others, implies that there is far more to human beings than we perceive through our brains and neurological apparatus. The world and our four-dimensional Selves are deeply connected to other levels and realms of reality; it is as possible to travel in and out of these psychic spaces as it is to live in our everyday world. Indeed, humanity is enthralled with the idea of both inner and outer space. We dream of exploring both realms, and the lure and adventure these frontiers of matter and consciousness represent fire our imagination. It has been more than 25 years since we first stepped upon the moon; humankind has always believed that we would one day do so. We regularly flock to the movies to see "Space 1999," "Space, the Last Frontier," and watch TV's "Deep Space Nine" and "Star Trek"; we seek "to go where no one has gone before." We are beginning to venture into the depths of our oceans. We dream of colonizing the moon and other planetary bodies in our solar system, perhaps even beyond. We yearn to explore distant galaxies and the far corners of the universe. More and more of us are also learning to travel to the deep, inner recesses of our psyches; we are beginning to explore the depths of our SOULS and of Consciousness Itself.

It's About Time

Exploration of space quite naturally leads us to ponder ideas about time. Science tells us that space and time form a continuum known as space-time and hence cannot be considered as separate dimensions. Essentially, we cannot speak of one without reference to the other, for where one goes the other is sure to follow. Time is the currency of life; we measure and record the events and experiences of our lives in the months, years, and decades of clock or *chronos* time. Most often, our perceptions of time are subjectively flavored and colored by our feelings, ideas, and circumstances so that it is almost impossible to separate ourselves from any experience of time. Space, time, and each of us are instrinsically bound and intertwined.

Moments in Time

At birth, we grow ourselves into time and space. Time accompanies every moment of our life's journey. Every pulse and every heart beat reflects the unidirectional flow we experience as time. Some describe a circle of time, without end or beginning, while others speak of a linear arrow of time. We imply that "time is money," when in fact, time is life. For time is the currency of our lives, just as the molecule A.T.P. (adenosine tri-phosphate) serves as the biological currency of energy exchange in the body. We cannot live without it. We necessarily exchange one moment of life for one moment in time, as we sequentially travel from the past, to the present, in order to meet our future. We experience chronos time as existing outside of ourselves, as the constant ticking of the clock that helps us measure the seconds, minutes, and hours that form our days, weeks, months, and years, and add up to centuries, millennia, and epochs. The clock ticks incessantly, as each of us grows from fertilized cell to fetus and on to experience our individual births. The instant of our birth is accorded great importance and significance; astrologers claim that the date and time of our birth contains an imprint or map of all the possible events that we may experience during our lifetimes. As we propel ourselves into a world of time and space, the relentless ticking of the clock continues to record our development from infant, to toddler, to pre-schooler, to pre-teen, to adolescent, to adult, and finally, to elder. Change accompanies and reflects our passage through time. As we develop and grow, each of us moves forward in time. With each rotation of the earth about our sun, we become another year older. We cannot go back to moments past. Yet, so many of us tenaciously cherish our memories of the past or, instead, rush to embrace a future we can sense only dimly. In so doing, we miss the present entirely, failing to live fully in the now. And each now is all we ever have, as we move from one now to the next now, and continually on to the next. ... German mystic Meister Eckehart expresses this idea, "There exists only the present instant. ... A now which always is itself new. ... There is no yesterday, nor any tomorrow, but only Now, as it was a thousand years ago and as it will be a thousand years hence."[23]

Our birth into time and space thus invites us to live our measure of days as best we can. When "our time is up," we will physically depart this space-time coordinate of existence and enter a state referred to as death. Many spiritual traditions teach that our consciousness returns to its SOURCE, to the realms outside of ordinary space and time, to a sacred space in which time knows neither beginning nor end.

Time is a subject of enormous complexity and abstraction, far beyond the thrust and scope of this book, which contends that Self and Psyche dwell not only within physical space-time, but are also deeply rooted and connected to hidden realms beyond the known world of space-time and matter. Only a brief and limited review of our understanding and perceptions of time is offered in support of this premise; the interested reader can refer to the bibliography for more in-depth explanations.

We experience time in numerous ways, and ascribe various qualities to our experience of a particular moment. Subjectively, some moments are precious, others are dreadful; and some moments disappear as we get "lost in time," while others drag and hang on endlessly. We rationally and objectively measure the length and quality of our lives according to chronos or clock time, whereas its meaning and essence can only be gleaned through an inner, irrational, non-linear understanding of time. Poets and mystics alike infuse their verses with sacred time, alluding to an eternal, enduring present that exists always. Altered states of consciousness, whether drug-induced or experienced in meditation, in reverie or in dreams, alter our perception of time and space. Our normal boundaries and conceptions of time and Self dissolve and we experience all of time as occurring simultaneously, and as enfolded in the spacious present. D. H. Lawrence shares his experience of such an awareness:

> *"'before' and 'after' were folded together, all was contained in oneness... and the beginning and the end were one. ... This was all, this was everything."*[24]

A similar notion is expressed by Wilber, "In this now resides the cosmos, with all the time and space in the world."[25]

Immersed in sacred time and space, we may experience timelessness or feel as if time has stopped altogether. We suddenly feel at one with everyone and everything around us, and experience our rightness in Universe and our connection with the ground of all being. These are profound moments of ecstasy, blessing, and grace. Each moment, big and small, perceptible and imperceptible, shapes and forms our lives and our awareness of time and space. As we record our lives along the continuum of time, we gain perspective, direction, and meaning. German philosopher Heidegger remarks that, "An individual is immersed in the flow of time, which provides both unity and a context in which to understand our unfolding human experience."[26]

Every community and nation projects a world-view... a paradigm or model incorporating their beliefs, perceptions, and experiences of space and time, of personhood, and of the cosmos. Such paradigms define, as well as offer, a rich contextual structure through which one gains a sense of identity and comes to understand the Self. One's perception of the world is determined by his or her values and beliefs. Scientists Ilya Prigogine and Isabelle Stengers reflect this idea in their observation that,

> "Each age searches for its own models of nature.
> For classical science it was the clock;
> for nineteenth-century science, the period of the Industrial Revolution, it was an engine running down.
> What will the symbol be for us? ... [Perhaps] a junction between stillness and motion, time arrested and time passing."[27]

Different Ages, Different Stages

Now that we have looked at humanity's scientific understanding of time and space, let us consider how our conceptions structure and influence our way of life. Three different paradigms experienced during three separate epochs of human history impart a unique sense of Self, a special relationship with nature, and a different understanding of humanity's place within the cosmos. A fourth and future paradigmatic stage remains to be experienced.

Each era represents a distinct stage of psycho-social development, as well as a particular world-view of reality during humankind's sojourn upon our planet. *Communion* reflects humankind's childhood. It is a time when First People inhabited earth and lived in harmony with Mother Earth. *Separation* symbolizes humanity's adolescence, as well as its quest for a new identity and Selfhood that would distinguish and separate our species from the rest of nature. *Chaos and Uncertainty* reflects humankind's present state of chaos and despair.

It is a time of crisis as well as opportunity. A fourth and future era, *Union and Reconciliation,* considers humanity's evolutionary potential and future. Our species, like many others before, faces disaster, devastation, and even the possibility of extinction. Yet, humankind also carries the seeds and potential to evolve to still higher levels of order, understanding, and development. It is we who will decide.

The choice is ours. Undeniably, each paradigm or model generates a world-view which humankind projects and ultimately experiences as "reality." Each paradigm creates its own world of experiences.

Communion

The first stage, Communion, symbolizes the instinctual, child-like trust and innocence of the First People* who awakened to life in the home of Mother Earth and Father Sky. Most early people enjoyed the generous bounty of nature and strove to learn her laws and cycles so as to live in harmony and balance with their mother. There was no need to question her rules, as humankind was protected, nurtured, and content. The cycles of life were intricately wed to the rhythms and rituals of each day, and we understood ourselves as part of a world that included all species of consciousness upon earth. All were humanity's brothers and sisters and all were respected. First People moved easily between inner and outer realms of reality and understood the unity and oneness of all life. Let us see how First People's conception of a sacred circle of time reflected a special and privileged childhood for humankind and the elaboration of a deep wisdom and understanding of the ways of nature.

First People are those ancient and indigenous peoples who first walked upon our planet. Some still do. They are the first of earth's children. Earthlings dwell in sacred space, a place where human beings experience a manifestation of divine power, and a place in which they sense their relatedness with the entire universe. Earthlings experience no barrier between inner and outer worlds, as both realities exist side by side. Both are known and both are explored. In the Kayapó universe in Amazonia, Brazil, "Experiences are precious personal conduits to the vast, dimensionless Cosmic Time"[28] that underlies and lends order in the Kayapó universe. The presence of spirit is everywhen and everywhere known and felt. The creator joins the created and becomes an integral part of daily life... and there is Communion!

---

*The world's indigenous peoples have collectively chosen to call themselves "First Peoples." *Once Upon ASOUL* slightly departs from this convention and instead refers to earth's original inhabitants as "First People," in order to distinguish humankind's ancient ancestors from their present-day descendants.

First People live upon Mother Earth and dwell beneath Father Sky (or Father Sun). They live in harmony with the laws of nature and trust in Earth's bounty. Occasionally, Earthlings experience the capriciousness of an angry, vengeful Mother Nature, expressed in storms, quakes, or floods. Yet they have faith that Mother Earth and Father Sky will regain their stability and calm will again prevail. It has always been so.

First People take only what they need and never waste or destroy anything. They desire to leave the world as they found it, and practice a balanced ecology that reveals a deep understanding of nature and a profound wisdom. Life is precious. An animal provides food, clothing, and shelter, and must be respected and honored for relinquishing its life in order that First People may live.

First People consider all beings upon the earth important; all are needed by the creator and all are interrelated. There is no hierarchy of consciousness, for all forms of life upon our planet pulsate with the Sacred Consciousness of the one creator. Animals are brothers and plant life are sisters. Natives hug trees and speak to rocks. Although there is an awareness that each species has its own unique way of apprehending the world, there is equality among all species. Each knows its place and each has purpose.

Many indigenous or "First People" perceive time to be a circle-spiral, weaving cosmic and earthly cycles of nature into their myths, rites, ecology, and the entire fabric of their lives. In this manner, they remain connected to their ancestors. Earthlings leave a timeless, priceless legacy of wisdom to their children, and a prescribed way of living in the world that has evolved over millennia. Hence, every generation becomes part of the eternal, timeless circle of life, living as those who have come before and those who will follow. When one dies, his or her bones and skull join those of the ancestors, and his or her presence continues to be felt and honored by all First People. Essentially, one lives on forever and achieves an immortality desired, but unknown, in our modern world.

Moreover, each person is needed and each has a place and role within the tribe or community. There are no wasted, discarded, or homeless people. The community bestows an all-encompassing identity upon its members that transcends time and space. Such an identity imparts dignity, self-esteem, and self-confidence and a way of life to First People. They recognize and know the One We Are.

Their days and years reflect the enduring seasons and the solstice, the migratory patterns of birds, fish, and fowl, and the timeless, biological patterns of nature. Earthlings recognize the interplay of night and day, and the patterns of sleep and wakefulness. They do not hurry time, nor do they live by the clock. They understand that time comes from within, even though they may, at times, experience it outside. Rather, First People follow the rhythms of the cosmos, in order to become one with nature. It has always been so. Ecclesiastes: 3 expresses a similar understanding of nature's rhythm, "To everything is a season, and a time to every purpose under the heaven."[29] The Kayapó of Brazil dwell in "the cyclical movement of time,"[30] "aware of the non-lineal realm of dynamic power that unifies all time and space."[31]

In sum, all First People respect and abide by the laws of nature, striving to live in harmony with all forms of consciousness. Universe is considered to be alive and aware. First People dwell in a world without beginning or end, aware that all life is sacred. They are privileged to walk among the Gods and to participate in the sacred mystery of life. It has always been so.

Separation

Gradually, humanity learned the rhythms and ways of earth, developed agriculture, and domesticated most of the animal kingdom. We were beginning to experience a new confidence and a growing sense of beingness that awaited expression and actualization. We grew impatient and arrogant since we had come to believe that our species was superior to other life forms. Within our boney cranium, our left brains began to stir and throb, and strange neurotransmitters and neurohormones began to flow. At first we hardly noticed when our frontal cortex began to look at things more rationally and logically. But soon we found ourselves questioning the very ways of nature and, defiantly, we began to believe that we could control and manipulate matter and impose our own rules and needs upon earth. We arrogantly considered that we were as smart as nature, maybe even smarter. We were filled with intense excitement and possibility. Humanity was experiencing a sudden, quantum shift in consciousness and it had taken us by surprise!

At some point in time, humanity made the decision to leave the sacred circle of life in order to establish an identity apart from nature. We meant no harm. Our species had simply become aware that we could step outside of ourselves and observe our actions in the physical world. Earthlings began to wonder if we could manipulate space and time and matter in ways not previously attempted, and were overjoyed that we could follow our own thoughts and ideas out into the world of matter. Humankind set forth to develop and explore a different sense of Selfhood, one rooted in the rational, linear, and logical mode of being. We proceeded to map new psychic territories of the Self. As adolescents seeking our own identity, humanity began a process of separation and individuation from all it had previously known.

Classical Science

As humankind separated itself from the rest of nature and sought to control matter for its own purposes, it developed another scientific paradigm with an entirely different understanding of time, space, and existence. We founded a Classical system of science, whose symbols were first the clock, and later, an engine that eventually would run down due to an increase in disorder and heat (entropy). We considered everything in space and time as discrete, separate objects, all capable of measurement and reduction into smaller components and parts. Things were now connected by a linear, sequential, causal chain of events that led to predictable effects and outcomes. We had discovered an objective, rational, mechanistic, deterministic, and predictable world. Our scientists were unraveling the mysteries of the universe and discovering its laws and properties. Soon, we believed, humanity would be able to control nature. Humankind was ecstatic.

Our physicists were telling us that time, like space, begins at the explosive instant of the Big Bang. It is therefore meaningless to consider any "before" prior to this enormous cosmic event, which gave birth to space, matter, and time, and all that we presently know. St. Augustine similarly reflects that our world was created, "with time, not in time."[32]

Indeed, our present Newtonian world-view envisions time as a linear, unidirectional arrow that sequentially travels from the past to the present to the future. Astrophysicist Stephen Hawking states, "The increase of disorder or entropy with time is one example of what is called our arrow of time, something that distinguishes the past from the future, giving a direction to time."[33]

Yet our arrow of time does not move, it simply points us toward "the future." We are the ones who endlessly and continuously move sequentially along the linear continuum of time, growing older in the process. Hence, "past" and "future" connote a relationship between one moment and the next, while recording our lives along time's flow provides perspective, direction, and meaning.

The consequences of the Classical-Newtonian world-view have been considerable. In order to extend its abilities and capacities through time and space, humankind chose to expand the left hemisphere potential of the human brain. We developed the rational brain, frequently ignoring the emotional aspects of humanity's nature. We amassed quantities of material goods and eagerly became homo-consumers, often losing the quality of our lives in the process. We chose rationality and science over humanity and humankindness. We developed an awesome technology that provided ease and comfort for many peoples of the earth, and we created communications systems that spanned and united the globe. We invented automobiles, airplanes, submarines, telephones, washing machines and dryers, air-conditioning, computers, photocopy and fax machines, and developed hydro-electrical power and a megaton arsenal of nuclear weapons, to name just a few of our accomplishments. Yes, we had stepped outside the sacred circle of time and explored a different kind of Selfhood, and we were quite proud of ourselves and our impressive achievements.

Nevertheless, we were also discovering a darker side to our technology and prowess over Mother Nature. As we separated ourselves, not only from nature, but also from all other species, we devised a hierarchy of value, divinity, and privilege, and we placed humanity upon its apex. We were becoming so powerful that many believed that our Gods never existed, while still others thought perhaps humankind was the heir apparent to fill this role in the world. Our arrogance had reached its peak, and we were regularly exploiting First People, rainforests, and nations. We were quickly using up nature's bounty and stores for our future, polluting the skys and waterways, experiencing soil-erosion and advancing deserts in our wake. We had been "fruitful and multiplied" and now humanity was overpopulating the earth and many were homeless and hungry. We began to experience more famines and natural disasters, and under the reign of "old paradigm science," we participated in war on a global scale, war that involved all of humanity.

Our inner cities have become slums and ghettos filled with despair, hopelessness, poverty, and never-ending violence. Our young have joined gangs and willingly die for any cause, rather than live meaningless, desperate existences. We, like the discrete objects we coldly and detachedly study in the laboratory, have become separated and isolated from nature. Many of our families are either single-parent homes or two-parent nuclear families without support systems. Children spend most of their waking hours in day-care centers until either one or two parents take them home. Our parenting skills are reaching zero, and our governments and economies can no longer provide for our overgrown populations. Like Mother Nature, we have lost our bounty and our stores for the future, and many are no longer able to sustain themselves, even in the present. Child abuse and substance abuse have become partners and live hand-in-hand in too many of our homes and communities. Indeed, our selves, homes, and communities very much resemble our Newtonian world-view of discrete, independent objects living mechanistically and deterministically in a rational, orderly world. Clearly we have lost our balance and our way.

Chaos and Uncertainty

"To be, or not to be, that is the question."[34]
—Shakespeare

"Here begins the new life."[35]
—Dante

Presently, humankind is in the midst of its third psycho-social stage of development. It is one of great promise, as well as one of serious psycho-social, economic, political, environmental, and spiritual crises. Humankind presently faces extraordinarily difficult problems and unprecedented challenges in almost every aspect and arena of life. Despite our brilliant technology and numerous accomplishments, humankind exists on the verge of global and spiritual crisis.

Everywhere a Quark, Quark

Many millennia ago, the Chinese recognized that within any crisis at least two different states existed side by side. They understood that every crisis is actually a moment of probability and uncertainty, with one probability leading to danger and possible disaster, the other espousing solution and resolution. Hence, one can readily see that the Chinese idea of crisis both predates and anticipates Western science's recent theories of uncertainty, indeterminism, chaos, and the science of complexity. Surprisingly, some of our hope and direction for restoring balance and harmony in our world may come from these newly disclosed findings of New Paradigm Science and Quantum Mechanics (Q.M.)

The laws of Newtonian science do not apply exclusively at subatomic levels of nature, nor in the far corners of the galaxies and the universe. Consequently, theories of Q.M. are challenging our notions of space and time, and in the process, are profoundly revolutionizing our understanding of nature and cosmos. At the boundary between the seen and the unseen, quantum physicists have uncovered a myriad of subatomic particles, all with strange names and bizarre properties. These include mesons, baryons, quarks, leptons, positrons, antiparticles, fermions, hadrons, pions, gauge bosons, and the newly discovered Bose-Einstein condensate.[36] Collisions of such particles leave traces and footprints which reveal their presence and their roles as the essential components of matter. Existence of all six flavors of quarks (Charm, Strange, Bottom [Beauty], Up, Down, and Top

[Truth]) has actually been verified.[37] It seems that quarks are everywhere.* They combine with six types of leptons (the electron, muon, tau, and three kinds of neutrinos) to form all matter. Without quarks and leptons there would be no matter and no universe. There would be no you and no me, for each of us is essentially combinations of quarks and leptons, held together by three known forces and a proposed "fourth" force. The fortunate scientist who succeeds in validating the existence of this fourth force will help humankind create a "theory of everything" (and earn a Nobel Prize as well). This theoretical fourth force has many names, including the Higg's boson, a.k.a. the Higg's particle or field, and recently, "The God Particle."[38] Science is seeking nothing less than a unitary theory of nature resembling the spiritual traditions' search for God and the source of ALL THAT IS.

Scientists tell us that we cannot describe space and time as qualitatively different dimensions. Rather, both are facets of a greater whole, and both are so intimately interconnected that they form a four-dimensional space-time continuum or unity. This unity gives rise to gravity, whose action curves and enfolds space-time around matter, ensuring that where one goes, the other must follow. Matter and space-time always travel together.

Consequently, in curved space-time, earth's orbit is no longer considered elliptical; rather, it is a helix. Although the earth returns to essentially the same space at the end of each rotation about the sun, it does not return to the same moment in time. We and everything upon our world have become a year older. Note that the genetic code for life embodied within the DNA molecule also assumes the structure of a double helix, suggesting that a helical pattern may carry some significance in nature. Einstein's Special Theory of Relativity recognized that time and age are relative. The faster one approaches the speed of light, the slower one ages, a concept known as "time dilation." It seems that how we age and experience the passage of time depends upon our relative frame of reference, which informs us about "where the where" is in the universe. Another idea posited by Einstein implies, "'the past, present and future' are only illusions since time does not 'happen' bit by bit, or one moment at a time. Time, like space, is extended and stretched out. ... 'Time is simply there.'"[39]

Another consequence of our four-dimensional space-time universe is to preclude a "personal" now. Again, in Einstein's words, "A human being is part of the whole, called by us 'Universe'—a part limited in time and space. He experiences himself, his thoughts and feelings

---

*Scientists are beginning to suspect quarks are not the fundatmental, indivisible particles of the universe. They may also consist of still smaller unknown particles.

as something separated from the rest—a kind of optical delusion of consciousness."[40]

Echoes of this idea are expressed by the Zen master Dogen,

"It is believed by most that time passes; in actual fact, it stays where it is. This idea of passing may be called time, but it is an incorrect idea, for since one sees it only as passing, one cannot understand that it stays just where it is."[41]

Motion pictures offer a useful analogy of this concept. An illusion of motion, i.e. movement, is gained by projecting a roll of film containing a series of still pictures, at a speed too rapid for the human eye to follow. Although the film moves frame by frame, the images in the film do not. Similarly, it seems likely that consciousness moves and shifts its focus from moment to moment, creating the illusion of time passing.

Quantum theory further suggests that there is a whole continuum of "presents," some existing in the past and some existing in the future. In a sense, the future already exists. A similar thought is expressed in the poetry of T.S. Eliot: "Time present and time past are both perhaps present in time future, and time future contained in time past."[42]

Uncertainty

Many scientists infer the presence of multiple pasts, presents, and futures, implying the likelihood that humankind experiences but one probable time line embedded within an infinite array of possibility. A world of probability and chance is a direct consequence of Heisenberg's Principle of Uncertainty (indeterminism). This fundamental theorem of Q.M. states that an inherent randomness, fuzziness, and unpredictability exists at the subquantum level of nature. Heisenberg's Principle asserts that we can statistically describe the full spectrum of possible outcomes, although we can never predict which one of these outcomes will actually result. Additionally, The Observer Effect contends that through measurement and observation, consciousness interacts with the property of uncertainty to affect outcome. It seems that we alter the properties of nature merely by probing into it. In Heisenberg's own words, "The path comes into existence only when you see it."[43] Capra agrees, "What we see depends upon how we look."[44]

There is an inherent randomness and uncertainty built into the universe, allowing for many possibilities and many potential outcomes of any experience or event. In ways not yet understood, consciousness interacts with matter to influence and to determine the course of events in our world.

We return to Bohm, who speaks of time as, itself, an order of manifestation, suggesting that within any given period, "the whole of time might be enfolded,"[45] implying that we are simultaneously living in all points of time. Such an idea of non-temporality corresponds to the property of non-locality of space. Remarkably, science is moving closer and closer to the interface between matter and non-matter, the manifest and the unmanifest.

Similar ideas are expressed in Tibetan Buddhism, which suggests that our essence of being, our "Clear Light," dwells outside of time and space and is independent of mind. Yet "Clear Light" resides, as well, within all beings and knows all space and time. "Clear Light" chooses and experiences each incarnation, projecting Itself into relative time and space as part of the karmic cycle of reincarnation. The Buddhists imply that everything occurs within the pure mind of "Clear Light."

Thus, it seems that We are the holomovement as well as the parts contained within it. Gradually, we begin to recognize that the holomovement, the Implicate Order, and the ground of our being are ONE. We grow aware that each moment or second of time inherently contains its own holographic imprint of time past, time present, and time future. Essentially, each moment or slice of time, like the hologram, simultaneously encompasses all of time, suggesting that consciousness and "All" exist in a spacious now, an enduring present without end or beginning.

Complexity

A more recent field, Complexity, the science of order and chaos, is developing none-too-soon, for it conveys a warning to humankind and insight for living in a world of uncertainty which is moving steadily toward chaos. The science of Complexity studies dynamic non-linear, self-organizing systems in which the whole is greater than the sum of its parts. An abundance and diversity of such systems are observed throughout nature and include the weather, our economics, the environment, our brain, the human genome, and humanity itself. Complexity teaches us that order and chaos are but two sides of one underlying unity that strives for balance. Nature's compulsion for disorder (entropy) is equally matched by a powerful tendency toward higher levels of order, structure, and patterns. Dynamic self-organizing systems are adaptive; their complex patterns spontaneously form, change, learn, and evolve in response to their environment.

New paradigm science tell us that everything is interconnected and interrelated in the fabric of the universe. Hence, any change in X (big or small) will affect Y, often triggering a whole cascade or sequence of events and changes that ripple throughout a system. The "butterfly effect" is a classic example of the sensitivity of chaotic systems to their initial conditions. "If a butterfly flaps its wings in the Amazon rainforest, a month later the air disturbance created may cause a hurricane thousands of miles away in Texas."[46]

A recent earthquake underneath the ocean, off the coast of Japan, raised the possibility that it could trigger large tidal waves as far away as the west coast of the United States. Fortunately, the prevailing conditions did not favor this possibility, and a natural disaster was averted.

Even the smallest amount of uncertainty can produce enormous change that suddenly brings a self-organizing, dissipative system to the very edge of chaos, uncertainty, and unpredictability. At such a moment, the possibility exists of either collapsing back into lower order forms, or for the system to re-organize and evolve into new orders and forms, with each level acquiring new emergent properties. Yet, a general principle of orderliness seems to govern the behavior of such complex and open systems, imposing some order upon uncertainty. Again, note the striking resemblance of these principles to the ancient Chinese conceptualization of "Crisis."

Wondrous as these newly formulated concepts of new paradigm science may be, what can they offer a world overwhelmed with rage and poised to self-destruct? Violence, crime, anarchy, as well as poverty, hunger, and pollution are everywhere rising and escalating. We have seriously decimated our earth and polluted the air we breathe. Our economies, our religions, our educational and governmental institutions, and our families no longer seem to work, and no one seems to know how to stem the tide. Despair, anguish, and signs of chaos are everywhere.

We have become, like our science, independent, discrete, rational objects who find it difficult, if not impossible, to relate and connect to one another. We are single or nuclear families without any support systems. We fear intimacy and lack commitment. Alienated from our true, deeper nature and source, it is no wonder that we feel so isolated and alone. We have forgotten our roots and spirituality, as well as the deep mystery of life. For far too long, humankind has stepped outside the sacred circle, separating ourselves from nature and from our own humanity. Humankind has traveled along a linear arrow of time, moving ever forward, only to find itself in severe crises, and at the edge of chaos and uncertainty.

Did we somehow take a wrong turn—for indeed, we seem lost and confused. Surely, humankind meant no harm! We thought we could control nature by manipulating its laws to our own purposes. We never considered the harm and damage we might inflict upon others in order to achieve our ends. We believed that we could do anything, and arrogantly set forth to meet this challenge without regard to Mother Earth, Father Sky, or all the species of consciousness that seemed to be in our way. We disregarded the warnings of the world's First People, who had long and wisely lived in harmony and balance with nature. Sun Bear of the Chippewa echoes this wisdom, "I do not think that the measure of a civilization is how tall its buildings of concrete are, but rather how well its people have learned to relate to their environment and fellow man."[47]

But these recent scientific findings, as well as the ancient teachings of all spiritual traditions, tell us that our world and its views and ideas are but one of an infinite array of possibility for humankind. All encourage us to choose a different path in order to experience a different outcome and world. It is possible to alter our awareness, our views, and our values, and accordingly change ourselves, our families, our ecology and our environment, our psychology, our sciences, our spiritual understanding, and our relationship with Mother Earth. New paradigm science implies consciousness can interact with other dimensions of being in order to manifest a benevolent technology, knowledge, and skills on earth's behalf, rather than in defiance of the laws of nature. We can rejoin the sacred circle of life. We can learn the knowledge of the heart of First People, and we can restore balance and harmony once more.

Eighty-three religious, political, and scientific leaders have signed "An Appeal for Joint Commitment in Science and Religion," entitled, "Preserving and Cherishing the Earth" which states, "The Earth is the birthplace of our species and, so far as we know our only home. ... We are close to committing—many would argue we are already committing—what in our language is sometimes called Crimes against Creation."[48]

I am one of those who argue that we have already committed crimes against creation, for we have cut down the trees of our global forests for enterprise and gain, and without regard or concern for the animal and plant life that call the rainforest home. They have no place to go, and like our increasing numbers of homeless people, many plants, birds, insects, and animals are now homeless. Many can no longer survive without their natural habitats and support systems; indeed, many are becoming extinct. We have also polluted our air, our lakes, and our waterways, killing fish, birds, and flora as we upset the delicate balance of nature. Everywhere our deserts are advancing and our soil eroding. Many millions, young and old, starve daily as the land can no longer support their needs and provide for them. Yes, we have desecrated earth and the heavens, and we have committed numerous atrocities against nature. Is it any wonder that nature grows impatient, angry, and vindictive?

The sacred Mayan text, "Popul Vuh," reaches forward in time to ask of us,

"Heart of Sky, Heart of Earth,
give us our sign, our word,
as long as there is day, as long as there is light.
When it comes to the sowing, the dawning,
will it be a greening road, a greening path?"[49]

Union and Reconciliation

"The cure of rivers is not a question of rivers,
but of the human heart."[50]

—*Tanaka Shozo*

Many of us yearn, deep in our hearts, for a simpler age and an easier style of life, a time when Earthlings lived as the First Peoples, enjoying the bounty and rhythm of Mother Earth and Father Sky. We yearn to go back. But in our particular world and Universe, the unidirectional linear arrow of time, as well as the sacred circle, can only move forward from one now to the next now, and then on to the next. While it is not possible to return to earlier stages of development, nor to avoid the rapidly approaching edge of chaos, we can consider how and why we have come to this particular moment in our history and consider, as well, what we can possibly do to affect the outcome of change already in process.

What can we learn from science, old and new, that may prepare us for this propitious challenge? Clearly, we have learned that change is inevitable, continual, and endless. Humankind, like other complex systems, is continuously developing and evolving, and whenever possible, seeks higher forms of expression. Each new level and order opens to and explores newly emergent properties and capabilities. For nature delights in the creative reorganization and restructuring of atoms, molecules, cells, and systems in order to know and experience itself in myriad ways. It is also part of humanity's heritage; it is imprinted in our genomes. Indeed, change is natural. Everything from the tiniest subatomic particle to the massive suns and galaxies of our cosmos experience cycles of growth and destruction, matter and energy continually converting one to the other in an endless, creative cycle through a timeless universe.

E = MC² and the Cosmic Dance

It is at the interface of matter and energy, the manifest and the non-manifest, that science and spiritual traditions now meet. The mystic experiences the endless cosmic dance of impermanence and creation, in which everything is energy. Physicists confirm the views of the mystic, implying that mass and energy are essentially different forms and aspects of the same unity. The equation $E = MC^2$ defines the basic equivalence between these two forms, revealing that mass is a form of energy and energy is a form of mass. Although mass may be transformed into various other forms (thermal, electrical, and kinetic), its total amount of energy is never lost. It is always conserved. Particles of different mass and energy are now recognized as dynamic patterns and processes within a four-dimensional space-time world. Every particle—and hence, every human being—is a dynamic bundle of energy that participates in the endless cosmic processes of creation and destruction.

The Dance of Shiva

> *"The dance of Shiva is the dancing universe; the ceaseless flow of energy going through an infinite variety of patterns that melt into one another... a pulsating process of creation and destruction."*[51]

Not long ago, while vacationing in Colorado, I experienced the cosmic flow of energy behind our world of physicality. It was a profoundly moving experience, one that has left an indelible imprint upon my Soul.

I sat upon a fallen log overlooking the banks of a nearby creek. The autumn day was warm and sunny and I felt a gentle breeze move across my cheek. A few squirrels were scurrying to and fro, and the golden leaves of the aspens were gently falling into the water below. I quietly gazed at the scene before me, about to close my eyes in meditation. Suddenly, the earth shook and everything surrounding me was in motion. I felt the earth quake. An inner voice quickly advised me to still my mind and to watch as the tranquil scene transformed itself into one of pulsating energy. Although vibrating at a different frequency, each object still retained its basic identity as tree, water, leaf, log, squirrel, and ground. Everything had instantly been converted into dynamic patterns of energy... nothing appeared normal and nothing was still except time. All vibrated and danced. I was immersed in a beautiful, pulsating golden dance of flowing energy.

Only later did I realize that I had never observed my own physical appearance during this experience. Most probably, I would have also viewed myself as a distinct vibrating pattern of energy. Undeniably, it would have been a shock to my "senses" to perceive myself as energy, rather than the solid being I recognize as Joyce.

The experience lasted only seconds, but it was profound. My momentary view of the field of energy beneath our world of matter, shape, and form has deeply and indelibly etched itself upon my psyche. I have consciously danced with Universe and experienced the indestructible Source of Consciousness which flows through all of matter, time, and space, indeed through each of us.

The above encounter with a field of pulsating energy beneath the world of manifestation generated a sense of fear and concern for humanity's future. Humankind, too, has arrived at the interface of matter and energy. It is here that our species must meet to decide its destiny. I wonder whether our species will find its way through the chaos and uncertainty of the present, and whether Earth People will return to the sacred circle of life.

As dynamic, non-linear, self-organizing beings, humanity is poised at the edge of chaos. We have come to a fork in the road and we must choose either life or death for our species, and for many other species as well. We have separated ourselves from nature, disregarded her laws, and destroyed much of our habitat and biosphere. Time grows precipitously short. Mahatma Ghandi reminds us, "The earth has enough for every man's need, but not for every man's greed."[52]

Humankind is consciously dancing the eternal dance of matter and energy; it is the continual dance of creation and annihilation, of life and death. We must choose our path carefully; the countdown has already begun! We see nations already collapsing into the old forms of hatred, blame, rape, and war. They project their despair and alienation upon other ethnic or political groups, and they have also raped, polluted, and declared war upon our planet. They do so at their own peril.

One choice is to transform the mass-energy of chaos and complexity into the creation and sustainability of new non-linear systems that will benefit all life upon our planet. Each small change that we make toward healing ourselves, families, societies, and world will be amplified. Our collective efforts will make an enormous difference in humankind's chances for survival. It is possible to change and transform ourselves and our world.

The questions one must now ask are: Can we influence the direction and course of this change, or will humankind, like so many other species, face the threat of extinction? Can we, like other complex non-linear, self-regulating systems, evolve instead to a higher order and level of understanding with newly emergent properties, characteristics, and possibilities? Can we determine which attributes and traits will sustain life and provide a fruitful future for humankind to explore?

How can we get enough people to work together to make the necessary changes? Humankind is enfolded within an auspicious moment of possibility. Nature warns that if we do not mend our adolescent ways of separation and exploitation, we will not long survive. This is the meaning of chaos and uncertainty.

We are presently forging a new model and understanding of science. But there is no need to throw out the old models that still work. What is unfolding is an inclusive and expansive science embracing interconnectivity and interrelativity. Science is regaining its conscience and SOUL. Consciousness (awareness) without conscience, or conscience without consciousness does not work. Science must wed the two together, understanding each as part of the same unity to realize its fullest potential.

It is time to develop a compassionate science, knowing both conscious awareness and conscience. It is time to return a sense of mystery and awe to ourselves, and to our world. Without a spiritual science, there can be no survival. Humankind is called forth to explore a higher level of development, to know itself as a reflection of a sacred Universe, and to acknowledge its connection to ALL THAT IS.

Notes

1. Vine De Loria, Jr., *God is Red* (New York, Grosset and Dunlap, New York, 1973), p. 158.
2. Ibid, p. 159.
3. A. Yusuf Ali (Transl.), *The Holy Qur'an*: 2nd. Ed. (New York, American Trust Publications, 1977), Sura XXX, II-4, p. 1092.
4. Frank Waters, *Book of the Hopi* (New York, Penguin Books, 1984), pp. 1-3.
5. James Cowan, *Mysteries of the Dream-Time* (Bridport, Prism-Unity, 1990), p. 25.
6. Stephen Mitchell (Ed.), *Tao Te Ching* (New York, Harper-Perennial, New York, 1991), p. 52.
7. Paul Davies & John Gribbin, *The Matter Myth* (New York, Simon and Schuster/Touchstone, 1992), p. 122.
8. Ibid., p. 122.
9. Paul Davies, *The Mind of God* (New York, Simon and Schuster, 1992), p. 68.
10. Stephen Mitchell (Ed.) *Tao Te Ching* (New York, Harper-Perennial, 1991), p. 25.
11. John Stewart Bell, "Nonlocality in Physics and Psychology": An Interview with John Stewart Bell in *Psychological Perspectives* (Fall-Winter, 1988).
12. Michael Talbot, *The Holographic Universe* (New York, Harper-Collins Publishers, 1991), pp. 15, 41.
13. Tyron Edwards, *A Dictionary of Thought* (Detroit: F. B. Dickerson Co., 1901), p. 196.
14. Ronald S. Miller, *As Above So Below* (Los Angeles, Jeremy P. Tarcher, Inc. 1992), p. xi.
15. Sir John Woodroffe, *The Serpent Power* (New York, Dover, 1974), p. 22.
16. Sir Charles Eliot, *Japanese Buddhism* (New York, Barnes & Noble, 1969), pp. 109-110.

17. Alan Watts, *Tao: The Watercourse Way* (New York, Pantheon Books, 1975), p. 35.
18. Ronald S. Miller, *As Above So Below* (Los Angeles, Jeremy P. Tarcher, 1992), p. 71.
19. Robert Kirshner et al., Harvard University in *Astrophysical Journal*, September, 1994. and Michael J. Pierce et al., Indiana University, *Nature*, September 29, 1994. Reviewed in *Science News*, Vol. 146, No. 15, Oct. 8, 1994, pp. 232-234.
20. Peter O'Connor, *Understanding JUNG, Understanding Yourself* (New York, Paulist Press, 1985), p. 17.
21. Paul Davies, *The Cosmic Blueprint* (New York, Simon and Schuster, 1988), p. 162.
22. Carl Pribram, *Chop Wood, Carry Water*, Rick Fields with Peggy Taylor, Rex Weyler and Rick Ingrasci (Los Angeles, Jeremy P. Tarcher, Inc. 1984), p. 210.
23. Peter Russell, *The White Hole in Time* (New York, Harper-San Francisco, 1992), p. 124.
24. D. H. Lawrence, *The Rainbow* (New York, Penguin, 1949), pp. 204-205.
25. Ken Wilber, *No Boundary* (Boston, New Science Library, 1979), p. 69.
26. Carl S. Hale, "Time Dimensions and the Subjective Experience of Time" in *J. of Humanistic Psychology*, Vol. 33, No. 1, 1993, p. 90.
27. David Suzuki & Peter Knudtson, *Wisdom of the Elders* (New York, Bantam Books, 1992), p. 188.
28. Ibid., p. 208.
29. Rick Fields, Peggy Taylor, Rex Weyler, and Rick Ingrasci, *Chop Wood, Carry Water* (Los Angeles, Jeremy P. Tarcher, Inc. 1984), p. 90.
30. David Suzuki & Peter Knudtson, *Wisdom of the Elders* (New York, Bantam Books, 1992), p. 208.
31. Ibid., p. 208.
32. Paul Davies & John Gribbin, *The Matter Myth* (New York, Simon & Schuster/ Touchstone, 1992), p. 141.
33. David Suzuki & Peter Knudtson, *Wisdom of the Elders* (New York, Bantam Books, 1992), p. 175.
34. William Shakespeare, *Hamlet,* Act III, Sc. i, line 48 in Justin Kaplan (Ed.) *Bartlett's Familiar Quotations,* 16th Edition (Boston, Little, Brown & Company, Inc., 1992), p. 196.
35. Alighiere Dante, La Vita Nuova (1293) In Justin Kaplan (Ed.) *Bartlett's Familiar Quotations,* 16th Edition, (Boston, Little, Brown & Company, Inc. 1992), p. 124.
36A. "A New Particle of Matter Unveiled," *Science,* Vol. 270, p. 1902, Dec. 22, 1995.
36B. Philip Yam, "Bose Knows P.R." *Scientific American,* p. 16, February 1996.
37. I. Peterson, "Extra data bolster top quark discovery," *Science News,* Vol. 147, p. 149, March 11, 1995.
38. Leon Lederman with Dick Teresi, *The God Particle* (Boston, Houghton Mifflin Company, 1993), p. ix.
39. Paul Davies & John Gribbin, *The Matter Myth* (New York, Simon & Schuster/ Touchstone, 1992), p. 82.
40. Stanislav Grof, M.D. with Hal Zina Bennett, *The Holographic Mind* (New York, Harper-San Francisco, 1992), p. 90.
41. Dogen Zenji, Shobogenzo, in Fritjof Capra, *The Tao of Physics* (Berkeley, Shambhala, 1975), p. 186.
42. Stanislav Grof with Hal Zina Bennett, *The Holotropic Mind* (New York, Haper-San Francisco, 1992), p. 113.
43. Fred Alan Wolf, *The Eagle's Quest* (New York, Summit Books, 1991), p. 145.

44. Fritjof Capra & David Steindl-Rast with Thomas Matus, *Belonging to the Universe* (New York, Harper-San Francisco, 1991), p. 124.
45. Ken Wilber (Ed.), *The Holographic Paradigm and other Paradoxes* (Boulder, Shambhala, 1982), p. 62.
46. Ian Stewart, "The mathematics of chaos" in *The Dictionary of Science* (New York, Simon & Schuster, 1994), p. 113.
47. Rick Fields with Peggy Taylor, Rex Weyler and Rick Ingrasci *Chop Wood, Carry Water* (Los Angeles, Jeremy P. Tarcher, Inc., 1984), p. 220.
48. David Suzuki & Peter Knudtson, *Wisdom of the Elders* (New York, Bantam Books, 1992), p. 168.
49. Ibid., unnumbered page
50. Ronald S. Miller, *As Above So Below* (Los Angeles, Jeremy P. Tarcher, Inc., 1992), p. 271.
51. Fritjof Capra, *The Tao of Physics* (Berkeley, Shambhala, 1975), p. 244.
52. Ronald S. Miller, *As Above So Below* (Los Angeles, Jeremy P. Tarcher, Inc., 1992), p. 257.

Footnote 9: Curt Suplee, *"Quark as Basic Particle May Be in Dispute,"* Washington Post, February 1996).

Selected Bibliography

David Bohm, *Wholeness and the Implicate Order* (London, Routledge and Kegan Paul, 1980).
John Briggs, *FRACTALS, The Patterns of Chaos* (New York, Simon & Schuster, 1992).
Joseph Campbell with Bill Moyers, *The Power of Myth* (New York, Doubleday, 1988).
F. Capra, *The Web of Life* (New York, Anchor Books, 1996).
R. Cowen, *Repaired Hubble Finds Giant Black Hole* in Science News, Vol. 145, No. 3, June 4, 1994, pp. 356-7.
Sukie Colegrave, *Uniting Heaven & Earth* (Los Angeles, J. P. Tarcher, Inc., 1979).
Peter Coveney & Roger Highfield, *The Arrow of Time* (New York, Fawcett Columbine, 1990).
Francis Crick, *The Astonishing Hypothesis: The Scientific Search for the Soul* (New York, Charles Scribners' Sons, 1994).
Paul Davies, *The Mind of God* (Simon and Schuster, New York, 1992).
Paul Davies, *The Last Three Minutes* (New York, Basic Books, 1994).
Arthur J. Deikman, M. D. *The Observing Self* (Boston, Beacon Press, 1982).
R. P. Feynman, *Q.E.D.* (Princeton University Press, Princeton, 1985).
Norman Friedman, *Bridging Science and Spirit* (St. Louis, Living Lake Books, 1994).
Robert Gilmore, *Alice in Quantumland: An Allegory of Quantum Physics* (Copernicus, New York, 1995).
David Ray Griffin (Ed.) *The Reenchantment of Science* (Albany, State University of New York Press, 1988).
Nina Hall (Ed.) *Exploring Chaos* (New York, W. W. Norton & Company, 1991).
Stephen Hawking, *A Brief History of Time* (New York, Bantam Books, 1988).
John H. Holland, *Hidden Order* (Addison-Wesley, Reading, 1995).
George Johnson, *Fire in the Mind: Science, Faith and the Search for Order* (Alfred A. Knopf, New York, 1995).
Michio Kaku, *Hyperspace* (New York, Oxford University Press, 1994).
Stuart Kauffman, *At Home in the Universe: The Search for the Laws of Self-Organization and Complexity* (Oxford University Press, New York, 1995).

Stephen M. Kosslyn and Oliver Koenig, *Wet Mind: The New Cognitive Neuroscience* (Free-Press, New York, 1992).

Rita G. Lerner & George L. Trigg (Eds.) *Encyclopedia of Physics,* 2nd. Edition (New York, VCH Publishers, Inc., 1991).

Steven Levy, *Artificial Life* (New York, Pantheon Books, 1992).

David Lindley, *The End of Physics: The Myth of a Unified Theory* (New York, Basic Books, 1993).

David MacLagan, *Creation Myths* (New York, Thames and Hudson, 1977).

I. Peterson, *At Last, Evidence of the Top Quark* in Science News, Vol. 145, No. 18, April 30, 1994, p. 276.

Ilya Prigogine and Isabelle Stengers, *Order Out of Chaos* (London, Heinemann, 1984).

Swami Rama, Rudolph Ballentine, M.D., Swami Ajaya (Allan Weinstock, Ph. D.), *Yoga and Psychotherapy* (Glenview, Himalayan Institute, 1976).

Jane Roberts, *The Unknown Reality,* Two Vols. (New York, Prentice Hall Press, 1977-1979).

Alwyn Scott, *Stairway to the Mind* (Springer-Verlag, New York, 1995).

John R. Searle, *The Rediscovery of the Mind* (Cambridge, MIT Press, 1992).

Rupert Sheldrake, *A New Science of Life* (Los Angeles, J. P. Tarcher, Inc., 1981).

Todd Siler, *Breaking The Mind Barrier* (New York, Simon & Schuster/ Touchstone, 1990).

Michael Talbot, *The Holographic Universe* (New York, Harper-Collins, 1991), p. 15.

Frank J. Tipler, *The Physics of Immortality* (New York, Doubleday, 1994).

M. Mitchell Waldrop, *Complexity* (New York, Simon & Schuster, 1992).

Steven Weinberg, *The Discovery of Subatomic Particles* (W. H. Freeman & Co., New York, 1990).

John White, *The Meeting of Science and Spirit* (New York, Paragon House, 1990).

Ken Wilber, *The Spectrum of Consciousness* (Wheaton, Quest Book, 1989).

Gary Zukav, *The Dancing Wu Li Masters* (New York, William Morrow & Company, Inc. 1979).

The Matter of Science

We are leaving behind the objective and rational realm of the scientist, in order to enter the cognitive-emotional world of the psychologist. The scientist inhabits a world of the senses, of space-time, matter, and action, whereas the psychologist dwells within a more subtle dimension of mind, a subjective domain of ideas, thoughts, feelings, and emotions. Psychology straddles the interface between the manifest and the unmanifest. Notice that psychology's language is more abstract, its ideas less defined.

Slaying the Dragon

(From the *Library* of Apollodorus)

Section three

n this section, we shift our focus from the sciences to psychology. Science is proud of its rational and logical investigation into the properties and characteristics of matter. Psychology also considers itself a scientific and rational discipline, while choosing to explore the irrational and intuitive processes of the psyche. Psychology is developing and exploring its own new paradigms which complement and affirm the ideas of new paradigm science. In their quest to find the ultimate, basic building block of nature, scientists have uncovered ever tinier, invisible units of matter, which have brought them to the interface between matter and non-matter. The domain between the seen and the felt is one familiar to psychologists. It is at this interface that I believe the two disciplines may begin to understand one another.

Psychology, like the sciences, has begun to identify the components and particles underlying human nature and personhood. By dividing the psyche and personality into smaller and smaller facets, psychologists hope to understand who and what we are, just as scientists hope to find the ultimate unit of matter. The more psychology explores each facet of personality and each aspect of the psyche, the more particles and facets they find.

The Gods Return Home

Humanity has historically projected its various perceptions of "goodness" and "badness" outside of our Selves. We projected our ideas upon nature as devas, elves, and fairies, as well as the gods of thunder, lightning, and rain. Later, we personified our concepts into female and male deities, only to recognize that our reflections are contained within these images. As we grew in our understanding of Self, we collected our gods and goddesses into several pantheons and hierarchies, and imbued them with more and more human characteristics. We created powerful religions, myths, and stories around them, to teach succeeding generations about the mysteries of the universe and creation. In each generation, a few wise individuals recognize that these myths and stories also provide a map for personal and collective transformation.

Gradually, humankind united its many different gods and goddesses into the *One God* we worship in various traditions and ways. We changed our myths and no longer recognize the old gods and goddesses. One by one, we banished them from our world. Yet they continue to exist within our psyches, urging us to awaken and grow into a fuller sense of Selfhood, and to reconnect to the cosmos.

Carl Jung once suggested, "We do not believe in the reality of Olympus, so the ancient Greek gods live on for us today as symptoms. We no longer have the thunderbolts of Zeus, we have headaches. We no longer have the arrows of Eros, we have angina pains. We no longer have the divine ecstacy of Dionysus, we have addictive behavior. Even though we no longer recognize the gods, we experience their powerful forces."[1]

Others, like Jean Houston, suggest that our separation from the gods, goddesses, and myths of yesteryear is responsible for the psychopathology experienced by humankind throughout the modern world. She states, "Wounding becomes *sacred* when we are willing to release our old stories and to become the vehicles through which the new story may emerge into time. ... If we would only look far enough and deep enough, we would find our woundings have archetypal power. In uncovering their mythic base, we are challenged to a deeper life."[2]

By creating a personal mythology, we become the authors of our own stories, allowing our lives to unfold with dignity and purpose. As we reclaim ourselves and our pathways, we again dance with the gods and goddesses within our psyches.

Religions and storytellers were the first to recognize the presence and power of the gods and goddesses who guide and influence our behavior in the world. The Greek philosopher Plato described a realm of ideas and forms that gives rise to the physical world. Indeed, he suggests that the process of entelechy* generates the dynamic patterns and blueprints of possibility encoded within each of our psyches, allowing us to create the many aspects and facets of our inner and outer personalities. Jung encountered powerful and godlike archetypes deep within the collective unconscious of humankind, such as Father, Mother, King, Warrior, and Magician. More recently, psychologists have described numerous sub-personalities and inner children residing within the level of the *personal unconscious*. Disowned and split off aspects of personality often remain dormant and hidden from Consciousness unless stirred and aroused into action by special circumstances. Once awakened, these facets of personality can autonomously act out their dramas and interfere with our lives until we acknowledge and heal them.

Amazingly, many of these wounded and repressed parts of our personality bear a striking resemblance to the gods and goddesses of

---

*Entelechy is realization or actuality as opposed to a potentiality; a vital agent or force directing growth and life.

earlier myths. The gods of Thunder and War, Thor and Mars, are now personified as the angry, wounded inner children existing within our personal unconscious. "Iron John, The Wild Man," and "The Women Who Run with The Wolves" are archetypes that cry out for discovery, inviting us to become authentic men and women rather than the prevailing stereotypes exemplified by the controlled, emotionless Marlboro Man, the macho, raging Rambo, or the sexually inviting female ingenue. We confront the wounding of Aphrodite and Demeter as the sexually abused inner child. Jennifer and Roger Woolger aptly describe the weakened, devalued goddesses within, "Aphrodite is ashamed of her sexuality; Athena questions her own ability to think; Hera doubts her own power; Demeter mistrusts her fertility; Persephone denies her visions; Artemis misunderstands her instinctive bodily wisdom."[3] We have also repressed and disowned magical, creative, loving, and sacred inner children.

By acknowledging and reclaiming our projections of goodness and badness, we open to healing and growth. The processes of healing and growth require reconciliation of the opposites and the polarities within our psyches, both personally and collectively. As we begin to honor and integrate the various demons, forces, gods and goddesses, archetypes, blueprints, sub-personalities, energy patterns, inner Selves, and potpourri of inner particles of personhood within each of our psyches, we release and make available creative forces that enable us to expand our understanding of Self and to live more satisfying lives. As we uncover and explore our multi-faceted, multi-leveled personhood, we awaken to a still greater consciousness, and we become whole. The gods and goddesses have returned home and are dwelling within each of us.

Ontogeny Recapitulates Phylogeny: The Cycle of Life

All that exist have a life cycle. The human psyche and our entire species experience organizing principles that direct and guide development and growth through our cycles of life. Every stage requires the individual and/or the species to complete certain psychological tasks in order to achieve new abilities and understandings. Thus, each succeeding phase incorporates and builds upon the abilities gained in earlier stages, offering additional characteristics and skills unique to its particular level of organization. This notion parallels similar ideas posited by Complexity and Chaos theories, and enables us to acknowledge that Psyche undergoes many evolutionary stages. Each phase also requires completion of specific developmental tasks, in order to psychologically mature and grow. If we fail to master a particular task, or if we experience abuse and ridicule at any stage, psychological development and potential are severely limited and compromised. Our drive will be to compensate and overcome these psychological handicaps throughout the remainder of our lives.

Essentially, the psychic development of every child recapitulates and repeats the psychological development of the species. This is the meaning of the phrase, "Ontogeny recapitulates phylogeny," a process which includes psychological, as well as biological and spiritual aspects of development, individually and collectively. Just as a child must learn to crawl before standing, and to stand before walking, the human psyche must experience different stages of development and psychological imprinting as it moves toward maturity and wholeness.

The Evolution of the Psyche

The human psyche is governed and directed by various organizing principles or archetypes. Shifts in governing principles, throughout individual and collective life cycles guide development from infancy through old age. The Matriarch or Great Goddess archetype operates during pre-conscious states of awareness, corresponding to infancy/early childhood and Communion. Under the Matriarch's influence, we experienced a pre-conscious Golden Age where all was complementary and relatively undifferentiated. "Things happen according to highly structured and pre-determined laws. ... Nature, not man or woman, is the authority. Instinct, not conscious knowledge, the guide."[4]

When Goddess rules, the child is identified with Mother. Indeed, they are one. Symbiosis of Mother and child in infancy and childhood (or Communion) is a period in which the dependent, vulnerable child (or species) is protected and nurtured by Mother or the "Mothering One." The child has many developmental tasks to complete, including bonding and attachment, then separation and the differentiation of "I" or Self from Non-Self. The new Self strives to grow itself outwardly into space and time.

The masculine (reality) and feminine (intuitive) principles of Psyche already exist within Matriarchal consciousness, but are not yet differentiated or separated. These two principles exist in all psyches. They refer to the ways we perceive and organize our experiences. They do not indicate gender. Neumann suggests: "Matriarchy and patriarchy are psychic states which are characterised by different developments of the conscious and unconscious, and especially by different attitudes of the one toward the other. Matriarchy not only signifies the dominance of the Great Mother archetype, but in a general way, a total psychic situation in which the unconscious (and the feminine) are dominant, and consciousness (and the masculine) have not yet reached self reliance and independence."[5]

Mythological stories of the Great Fall, or expulsion from the Garden of Eden, symbolize the overthrow of Matriarchal consciousness, as well as the separation within Psyche of the masculine and feminine principles. Such stories epitomize a shift from pre-conscious (pre-personal) unity to polarization, differentiation, and expanded awareness. The masculine is required to usurp the authority of the Goddess, and to "slay the dragon." It is the masculine that urges us toward autonomy and independence through separation and specialization. This psychic event has occurred again and again throughout the history of the human species. The task of *separation and individuation* is one required of every individual, and is so important that it reappears during several stages of the life cycle. Each time we encounter this challenge, we must understand and master it anew. Although various spiritual and psychological disciplines understand and interpret this psychic task differently, all address it. The influence and imprint of the patriarch or masculine principle correspond to the epoch of Separation, described below.

The Crisis of Oedipus

Freud, Steiner, Jung, and many of their followers consider the myth of Oedipus central to their different interpretations of this critical stage of intrapsychic development. Freud contends that every child must give up the instinctual (id) desire for an incestuous relationship with Mother. Next, the child identifies with the parent of the same sex, and assumes the roles, attitudes, and restraints appropriate to this gender, allowing the superego to emerge. Resolution enables a child to grow toward puberty, adulthood, and to eventually embrace mature love and marriage.

Jungians symbolically interpret this universal, archetypal myth as the child's need to "slay the dragon," thereby overthrowing the powerful grip of the Goddess or Matriarchal consciousness which engulfs his or her psyche. At this very special moment, the psychological unfolding and flowering of a *separate* individual begins. Hence, the Oedipal conflict is an auspicious task, offering each boy and girl psychological birth and personhood. It is the first time that the Self asks the timeless question, "Who am I?"

Neumann, a Jungian analyst, contends that Oedipus chose to slay his father in order to prolong his incestuous relationship with his mother. Neumann further suggests this "parracide" is psychic and symbolic. The biological father is not destroyed, rather it is the masculine principle that is disavowed. Oedipus kills the masculine principle in order to return to the earlier pre-egoic state of oneness with the Great Mother. The price exacted for this psychic regression is blindness. Oedipus failed to recognize (see) that a state of consciousness which is helpful, necessary, and appropriate for an earlier stage of psychological development causes illness and pathology at another. Oedipus regressed rather than progressed. The essential warning contained in the myth of Oedipus tells us that the spiral path of growth and awareness always urges humankind forward.

Rudolph Steiner imparts an additional spiritual dimension to this psychological drama. He suggests that Soul, as well as Psyche, includes both masculine and feminine principles. Our task is to harmonize the two principles in order to realize wisdom. Steiner considers the process of separation into masculine and feminine components a necessary prerequisite for later spiritual growth. Moreover, Steiner believes the rise of Christianity signals an important shift in human awareness. For Steiner, the physical realization of Christ represents an evolution in consciousness, one that makes it possible for such complementary principles to unite and reconcile.

The comparable drama for young girls, the Electra Complex, is far more demanding, because the young girl must psychologically separate from Mother while still continuing to sexually identify with her. She must also renounce any sexual desire and love for Father and, as an adult, ultimately seek fulfillment of these drives from another man.

Regardless of their different approaches and interpretations, each psychological school acknowledges the profound, transformative implications of this particular psychic drama. All believe that this myth speaks of a critical time of *separation* from Mother. Each recognizes this stage as leading to the emergence of individual (Ego) consciousness. In truth, the Oedipal conflict is nothing less than a ritual of initiation into another state of awareness. The successful or unsuccessful resolution of the Oedipal crisis deeply affects and influences each person's sense of Self and determines his or her level of psychological functioning throughout life.

It appears that to find and to "know our Self," we have to first repress and hide parts of ourselves. Only after we form an "I," a "me," and a personal identity, can we complete the rest of our psychological and spiritual growth. Our psychic journey propels us onward toward the rediscovery and reclaiming of these inner cast-off, disowned, split off portions of ourselves, personally and collectively. In order to mature and grow, each of us is obliged to assume responsibility for our projections, feelings, ideas, and acts, so that we might reclaim, heal, and integrate these aspects of our psyches and once again become whole and unified.

Hence, the Great Mother imparts an oceanic experience of love and unity, a sense of individual value, and nurtures the vulnerable child until he or she can function reasonably well in the world. Gradually, the child must let go of its protective cocoon and oneness with Mother, and begin to make its own way in the physical world. The separation of the unified psyche of the Goddess into the masculine and feminine principles occurs around age three. This allows for the development and ascendence of the masculine principle, and for the emergence of Ego consciousness and individuality. Henceforth, one experiences the world of polarity and duality, of "I" versus "thou."

Without the overthrow of the matriarch, there can be neither separation nor development of the feminine and masculine principles, nor individualization, nor conscious choice. Unless the human psyche experiences the evolutionary process of separation, it cannot move beyond the passive protection of the matriarch. Without awareness of "I" and "thou," we can never fully enter into relationship. Rather, separation and the rule of the masculine set the stage for a more expansive consciousness to emerge which seeks diversity, desires free choice, and pursues a conscious collaboration with Mother Earth.

The Feminine Principle

Presently, Psyche, individually and collectively, is correcting its lopsided view of Self and world. To do so, it must embrace its "opposite," the feminine principle. The feminine restores balance and brings wholeness to our inner and outer worlds. Many longing to experience the peacefulness, nurturance, and protection of an earlier Golden Age urge us to re-embrace the Goddess. They wish us to return to simpler ways of seeing and knowing the world. But in our physical reality, we cannot move backward in time or development without serious psychological regression and consequence. It would be like going from walking to crawling. We are urged ever forward. In so doing, we grow a year older and hopefully, wiser and more compassionate.

Unfortunately, some groups have confused the Goddess with the Feminine aspect of Psyche. They are not the same. The Goddess is essentially Earth Mother, who enfolds the child within her bosom so that child and mother are one. The Goddess rules over an undivided house. There is no need for separation or choice. All is provided. Throughout eternity, She remains the great nurturer, provider, and caretaker.

In contrast, the feminine serves as a bridge between worlds and realities; she is the gateway to transpersonal realms. She is trusting, warm, and receptive, always open to inner perception and wisdom. The feminine desires relationship with the masculine, moving both aspects toward unity and wholeness. The realization of the positive feminine leads to an expanded "I" and imparts great awareness and compassion.

Confirmation and support for this premise can be found in the myth of Demeter and Persephone* which (1) tells us that the matriarch and the feminine are two different organizing principles, each presiding over a different stage of psychic development, (2) warns us against regressing to an earlier stage of psychic development: In order to grow and evolve, Psyche must strive to reach higher levels of awareness, and (3) reveals that Feminine undergoes deep transformation in the silent depths of the unconscious.

Essentially, myths and fairy tales about women portray Feminine's journey as a passive withdrawal into the underworld, a silent suffering and waiting for deliverance, and an ultimate awakening into conscious union with the masculine.

Consequently, the awakened feminine heralds the possibility for spiritual birth and the reconciliation of opposites; she is the harbinger of a new evolutionary stage for humankind. Thus, Demeter's story speaks directly to the experience of all women, as well as to the urgent need for change in our present world.[2]

Acceptance of the differences between the Goddess and the Feminine frees humankind to embrace a more mature era of psychic development in which the masculine is balanced by the feminine. Through processes known as individuation and the Sacred Marriage, these two diverse principles unite. When the masculine and feminine aspects work in cooperative harmony, both the individual and the species experience unlimited creative expression. The integration of the two affords great healing ability, deep spiritual insight, and compassion. It is through the unification of these elements that we can, once again, realize harmony and peace upon earth. A global evolution in consciousness has begun, carrying with it the seeds of expanded Selfhood (refer to Tables 1 and 2).

---

*A brief version of Homer's myth of Demeter and Persephone, along with its significance, appears in Appendix I.

Table 1
Three Parallel Life Cycles

Human Species	Individual	Psyche
Communion	Infancy-Childhood	Goddess/Matriarch
Separation	Adolescence	Masculine/Reality
..........Chaos-Uncertainty..........		Feminine/Intuition
..........Reconciliation-Reunion..........		Sacred Marriage

Table 2
Psycho-Social Development of the Self

Psychological	Psyche	Consciousness
Dependent	Goddess	Pre-Personal
Independent	Patriarch	Personal
Interdependent	Feminine	Transpersonal
One	Sacred Marriage	Transcendental

The Process of Transformation

The process of transformation is central to all psycho-spiritual traditions. Ken Wilber, a founder of Transpersonal Psychology,* contends that every school of psychology and every spiritual tradition addresses a particular level of consciousness. These levels collectively create and span a "Spectrum of Consciousness."[7] Each discipline and level endeavors to increase awareness of projected, denied, and repressed elements of Psyche. Each encourages a person to re-claim, identify with, and integrate previously disowned, rejected parts of Self, so that he or she becomes more integrated. All psychological and spiritual disciplines attempt to heal the split, as well as to reconcile the opposites that exist on its particular level of expertise. As we become aware of Psyche's contents, we are no longer its victims and we are no longer out of control. We are empowered, our sense of identity is expanded, and we feel more solid. As we continue along our individual paths, new levels of consciousness invite and await our exploration.

Without adequate preparation, one is ill-advised to begin a spiritual journey. A strong ego is necessary before one can touch the Soul. Yet if ego is rigid and inflexible, growth cannot occur. Development of

---

*"Transpersonal Psychology explores stages of human development lying beyond the level of the fully developed ego." Washburn, p. 1.

an ego strong enough to relinquish its centrality yet able to serve the deeper aspects and goals of Psyche is our task. Ego is neither abandoned nor dissolved, only asked to release its tenacious hold upon Psyche, so that a larger, more sacred Selfhood may emerge. In this manner, all aspects of the SELF reconnect and cooperatively work together. We are no longer divided and fragmented. A Sacred Marriage has transpired.

Some conceptualize the personal unconscious as an aggregate or community of independent Selves and energy patterns, which we can learn to recognize and to acknowledge. Developmentally, the task of the Self is to weave all the conscious and unconscious aspects of Self and personality into a recognizable, cohesive whole that we refer to as "you," "I," and "me." Outwardly, I am experienced in many different ways and in various roles such as wife, mother, sister, daughter, teacher, psychologist, and scientist. I constantly shift from one role to another. Still discernable is a quality of Joyce-ness which blends these many facets and activities together. Similarly, the inward personality combines and incorporates many inner voices and particles of personhood. The inside-outside person is, indeed, multi-faceted. Paradoxically, we are the One and we are the Many.

New paradigm, Third Force, or Humanistic psychologies emphasize personal growth and self-actualization, while Fourth Force psychologies (Jungian, Depth, Sacred, and Transpersonal schools) address transpersonal, transhuman, and cosmic aspects of being. Third and Fourth schools of psychology already venture beyond ordinary levels of the ego, to explore existential, collective unconscious, and transpersonal realms of Selfhood. Humanistic and Transpersonal psychologies describe a multi-layered psyche which encompasses the collective unconscious, the transpersonal and the transcendental realms. Archetypes, deities, and demons are said to reside in these other domains. To this list, the spiritual traditions bring the SOUL.

Transpersonal psychologies also guide individuals who travel the Path of the Heart in search of self-realization and enlightenment. The Transpersonal psychologies invite Ego Self to awaken and embrace its greater SELF or SOUL. In the process, Ego Self evolves and becomes whole. Both Third and Fourth force psychologies and spiritual traditions acknowledge and explore transpersonal and transcendental realms. Traditional psychiatry and psychology, concerned primarily with ego states, perpetuate the view of spiritual experiences as pathological. The newer psychological schools, however, are already knowledgeable in healing the split between mind, body, and spirit.

Traditional psychiatry and psychology are also challenged to relinquish their ego-centric view of the psyche as essentially pathological and untrustworthy, in order to realize their greater potential. Failure to honor humankind's greater potential and creativity, individually and collectively, leads to frustration, apathy, and limited growth. Psychology stands at a unique moment of possibility. Genesis of a sacred and compassionate psychology, concerned with healing the Soul, affords humankind the guidance and assistance necessary to achieve this wholeness and peace.

As Above, So Below

> *"I live my life in growing orbits,*
> *which move out over the things of the world.*
> *Perhaps I can never achieve the last,*
> *but that will be my attempt.*
> *I am circling around God, around the ancient tower,*
> *and I have been circling for a thousand years.*
> *And I still don't know if I am a falcon,*
> *Or a storm, or a great song."*[8]
> —*Transl. by Robert Bly*

Inner and outer realities are intricately linked and interconnected on many levels and many dimensions within a circular-spiral matrix. Change and transformation in one dimension creates a corresponding echo of transformation in all others. The laws of nature and mathematics require symmetry. If we add an integer on one side of a mathematical equation, we must add its equivalent to the other side in order to maintain balance. It is, therefore, not surprising to observe reciprocity and symmetry at work in our psyches. As we gain awareness and wholeness in our inner realities, we experience awareness and wholeness in our outer world as well. Balance is achieved. Inner joy and gladness will spill rapturously into the material world, even if people surrounding us are angry and irritable. Although sensitive to the pain of others, we remain centered and continue to grow. Similarly, if we are inwardly sad and angry, we will automatically imbue our physical world with many shades and tones of sad and angry feelings, until we inwardly shift to other emotions and thoughts. As our inner consciousness expands and transcends one level of understanding to reach another, corresponding expansions and leaps are simultaneously experienced in our ordinary world. "As Above, So Below" is the essence of Jesus' teaching: "When you make the two one, and when you make the inside like the outside and outside like the inside, and the above like the below, and when you make the male and female one and the same... then you will enter [the Kingdom]."[9]

Different Dimensions, Different Perspectives

The interface between matter and non-matter, and the above and the below, is an extraordinary dimension. Here, infinite worlds, universes, and realities meet and converge, and the paths of three major disciplines—science, psychology, and spirituality—intersect.

A Word of Caution

Whether we speak of the world of matter or of an invisible realm of non-matter, both are inhabited by Psyche. We conclude that the body and Psyche are complex patterns of interacting particles and forces, existing on many levels, collectively creating a multi-faceted, multi-leveled, and multi-dimensional Selfhood and Universe.

Both psychological and spiritual traditions speak of non-physical dimensions of existence. Hence, it is not possible to weigh, measure, or define such dimensions in terms of the physical world. Why, then, do we insist upon trying to validate the hidden, invisible dimensions of Psyche and Soul with scientific methods designed for use in a material, physical universe? It doesn't work! Although there are many parallels, correspondences, and similarities among the fields of science, psychology, and spirituality, there are also incredible differences. Efforts to prove or disprove one by methods of another inevitably lead to frustration and conflict.

Experimental methodology may enable us to understand biochemical processes, or the manner in which time affects memory, but it is of little help when we address values, meaning, and the essence of life. Spiritual experience is individual, personal, and transformative. It is direct! No intermediaries are required. Perhaps that is why encounters with SPIRIT carry an indelible stamp of certainty and last a lifetime. Whenever we come face to face with SPIRIT, we are dealing with the very ground of our being, unbounded by space-time, and without duality or definition. It is no wonder that attempts to explain the mystical, transcendental realms only diminish and trivialize them. Authentic spirituality only happens on the other side of words. As we journey more deeply into the psyche, we find that our experiences can no longer be conveyed or translated through words. They can only be felt, experienced, realized.

Each of the three disciplines speaks and sees the world quite differently, science principally using left hemisphere, psychology emphasizing the right hemisphere, and spiritual traditions uniting both cerebral hemispheres. In this way, we obtain a more complete and balanced understanding of our world. In mystical and spiritual experiences, subject and object become one in the act of knowing, and duality ceases.

The fields of science, psychology, and spirituality address three different realms of the psyche, and do not deal with the same level of experience. Each higher level transcends and incorporates the characteristics and properties of the one below it, as well as adding unique and special attributes of its own. It is both additive and integrative. The lower realms are sub-sets of higher levels and include only certain characteristics of the domain above. Hence, caution is advisable whenever comparing one discipline with another.

A number of models encompassing from three to 20 different levels in the "Great Chain of Being" have been proposed by various spiritual traditions.* Each level or realm is complete within itself and each gives rise to a particular understanding of reality. We have already discussed these organizing principles of reality in terms of science and psychology and the evolution of the psyche. In Wilber's schema, taken from Eastern traditions, matter is the domain and realm of physics. The domain of life is concerned with biological life processes, while mind is explored by psychology. The SOUL is the interest of theology, while SPIRIT is the focus of the Mystery schools. Each higher level is both additive and integrative. Refer to Figure 3.

The Realms of Spirit, Section Five, explores other paradigms describing multiple dimensions of reality including the Four Worlds of Emanation, Creation, Formation, and Manifestation and the previously mentioned three worlds of the shaman (Upper and Lower Imaginal Realms, and the world of earthly existence). As in the Great Chain, each world is nested within the next, and each world symbolizes a specific level or grade of enlightenment and awareness.

Whenever mystical traditions speak of the Four Worlds of Emanation, Creation, Formation, and Manifestation, they not only describe different dimensions of Psyche and reality, they also reveal the processes of Consciousness Itself. In order to manifest change or action in the material world, consciousness is required to ascend and descend these four worlds of awareness.

The Worlds of Emanation, Creation, Formation, and Manifestation are none other than the transcendental, the transpersonal, existential, and ego-levels of consciousness of Transpersonal psychologies. They

---

*A simple version of the "Great Chain of Being" of Eastern mystical traditions is illustrated in Figure 3.

Figure 3
The Great Chain of Being

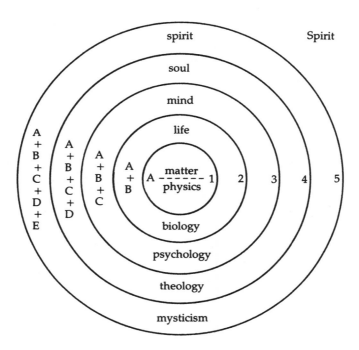

"Every level in the Great Chain transcends but includes its predecessor(s). The higher level does not violate the principles of the lower; it simply is not inclusively bound to or explainable by them. The higher transcends, but includes the lower and not vice-versa."

Source: Wilber[10]

are also the levels of (1) Soul and Spirit, (2) the collective unconscious of Fourth force psychologies, (3) the Humanistic psychologies, and (4) Psycho-analytic and Behavioral psychologies, respectively. Essentially, the transpersonal psychologies and mystical disciplines reveal similar hidden, invisible levels, dimensions, and realms of existence. Each shares its special understanding and unique perception of the awakening of Self to SELF. Refer to Table 3.

Table 3
Great Chain of Being—Worlds within Worlds

Gt.Chain	Consciousness	World	Discipline
Spirit	Universal	emanation	God/Source
spirit and soul	transcendental	creation	Spiritual Trads.
Mind/Psych.	transpersonal	formation	4th Force Psych.
Life/Biology	existential	manifestation—	Humanistic
Matter/Science	ego	and action	Ego/Behavioral/ Psycho-Analytic

Every person must re-trace this sacred pathway, moving from one world to the next, if he or she wishes to attain enlightenment and spiritual awareness. First, we heal the split on the ego-level and transcend to the existential level. Next, we transcend the existential to reach transpersonal and collective levels. Only a few attain transcendental, spiritual awareness. The path from Self to SELF moves from the world of manifestation, to the world of formation, on to the world of creation, and finally opens to the possibility of glimpsing the world of emanation.

Paradoxically, when one transcends into the realm of Consciousness Itself, one understands that there are no levels, no dimensions, no higher, no lower, no sacred, and no profane. All is ONE.

"We are what we think.
All that we are arises with our thoughts.
With our thoughts we make the world."[11]
—Buddha

The sacred traditions suggest that our present understanding of reality and consciousness needs to be expanded in order to more fully know and realize ourselves. While we cannot fully know the unknowable, it is possible to gradually approach some of these hidden dimensions, to interact with them, and thereby grow in awareness.

All spiritual traditions teach that the core SELF, our SOUL, is a timeless, conscious, and eternal process, rather than an object or impermanent being. Eternity is transformed as each of us participate in the endless rhythms of creation and annihilation. Each of us creates our world and our experiences, and each individual consciousness participates in the creation of the whole. Human beings are an intrinsic, vital aspect of Universe. It would be incomplete without us. In some fundamental sense, the Universe needs humanity in order to know itself. We are the ones who connect heaven and earth. It is humankind, standing at the edge of chaos, who must decide the world of the future, not only for ourselves, but also for the entire cosmos. We have been given an awesome responsibility and choice.

Separate and diverse principles within the human psyche strive for reconciliation. These elements are complementary, rather than opposites. An "either/or" position that views these aspects as opposites reflects old paradigm thinking. In truth, these aspects are interrelated and cooperative counterparts which, when united, offer a larger mosaic of understanding and consciousness. Whenever two or more diverse psychic elements join their perceptions, both individuals and species move from a stark black and white world, to one painted in exquisite rainbows of color. Union is additive as well as integrative. The world, as well as our understanding, is enhanced, and a richer repertoire of possibilities and options unfolds. We not only become more whole, we also get to see more of the whole.

I believe that a mature humanity will learn, once again, to live in accord with the laws of earth and of Universe, as well as to acknowledge the rights and needs of other species. We will use our knowledge and creativity for universal good, not for personal gain. Humanity has come to the threshold of another stage of awareness. Within this more mature consciousness is the potential for a more peaceful existence and a new renaissance for all species. A new evolutionary form of consciousness and personhood seeks to know and express itself in the world to be. Transpersonal specieshood welcomes us all.

Notes

1. Carl G. Jung, in *As Above, So Below* (Los Angeles, Jeremy P. Tarcher, Inc., 1992), p. 114
2. Jean Houston, *As Above, So Below* (Los Angeles, Jeremy P. Tarcher, Inc., 1992), p. 115.
3. Jennifer and Roger Woolger, *As Above, So Below* (Los Angeles, Jeremy P. Tarcher, Inc., 1992), p. 155.
4. Colegrave, S, *Uniting Heaven and Earth* (Los Angeles, Jeremy P. Tarcher, Inc., 1979), p. 40.
5. Neumann, E. *On The Moon and Matriarchal Consciousness in Fathers and Mothers* (Vitale et al.), pp. 40-41.
6. Funk and Wagnalls, *Standard Dictionary of Folklore, Mythology and Legend*. Maria Leach and Jerome Field (Eds.) Harper and Row, San Francisco, 1972, pp. 306–307.
7. Wilber, K., *The Spectrum of Consciousness* (Wheaton, The Theosophical Publishing House, 1989).
8. Bly, Robert, Transl. in J. Abrams (Ed.), *Reclaiming the Inner Child* (Los Angeles, Jeremy P. Tarcher, Inc., 1990).

9. Pagels, Elaine, *The Gnostic Gospel* (New York, Vintage Books, 1979), p. 155.
10. Wilber, K. (Ed.) Introduction: Of Shadows and Symbols in *Quantum Questions: Mystical Writings of the world's great physicists* (Boston, Shambhala, 1984), pp. 15-16.
11. Byrom, T., *The Dhammapada: The Sayings of the Buddha* (New York, Vintage, 1976).

Footnote 7: Michael Washburn, *Transpersonal Psychology in Psychoanalytic Perspective* (Ithaca, State University of New York Press, 1994).

Selected Bibliography

Alter, W. "The Yang Heart of Yin: On Women's Spiritual Nature" in *The Quest*, Winter, 1993 pp. 40-45.
Belenky, M. F., Clinchy, B., Goldberger, N. R., & Tarule, J. M. *Women's Way of Knowing: The Development of Self, Voice and Mind* (Basic Books, Inc., 1986).
Castaneda, C., *The Eagle's Gift* (New York, Simon & Schuster, 1981).
Castaneda, C., *The Fire From Within* (New York, Simon & Schuster, 1984).
Compton, M., *Archetypes of the Tree Of Life* (St. Paul, Llewellyn Publications, 1991).
Cowan, J., *Mysteries of The Dream-Time* (Bridport, Prism Press, 1990).
DeLoria, V., Jr., *God is Red* (New York, Grosset & Dunlap, 1973).
Edinger, E. F., *Anatomy of the Psyche: Alchemical Symbolism in Psychotherapy* (La Salle, Open Court, 1985).
Eisler, Riane, *The Chalice and the Blade: Our History, Our Future* (San Francisco, Harper & Row, 1987).
Eisler, R. & Loye D., *The Partnership Way* (San Francisco, Harper-Collins, 1990).
Fields, R., Taylor, P., Weyler, R & Ingrasci, R. *Chop Wood, Carry Water* (Los Angeles, Jeremy P. Tarcher, Inc. 1984).
Fox, M., *A Spirituality Named Compassion* (San Francisco, Harper & Row, 1990).
Gimbutas, M., *The Language of the Goddess* (San Francisco, Harper & Row, 1989).
Goleman, D., *The Meditative Mind* (Los Angeles, Jeremy P. Tarcher, Inc. 1988).
Griffin, D. R. (Ed.), *The Reenchantment of Science* (SUNY Press, Albany, 1988).
Halevi, Z., *Psychology & Kabbalah* (York Beach, Simon Weiser, Inc., 1987).
Halevi, Z,. *ADAM and the Kabbalistic Tree* (York Beach, Simon Weiser, Inc. 1990).
Halifax, J., Ph.D., *Shamanic Voices* (New York, E. P. Dutton, 1979).
James Hillman, *The Soul's Code: In Search of Character and Calling* (Random House, New York, 1996).
Harner, M., *The Way of The Shaman* (New York, Bantam Books, 1986).
Judith, A. & Vega, S., *The Sevenfold Journey* (Freedom, The Crossing Press, 1993).
Jung, C. G., *Memories, Dreams, Reflections* (New York, Pantheon Books, 1973).
Kaplan, A., *Jewish Mediation: A Practical Guide* (New York, Schocken Books, 1985).
Knight, G. & MacLean A., *Commentary of the Chymical Wedding* (Edinburgh Magnum Opus Hermetic Sourceworks, 1984).
Leonard, L. S., *On The Way To The Wedding* (Boston, Shambala, 1986).
Luke, Helen M., *The Way of Woman: Awakening the Perennial Feminine* (Doubleday, New York, 1995), pp. 103-120.
Maslow, A. H., *Toward a Psychology of Being* (New York, Van Norstrand Reinhold Company, 1968).

Maslow, A. H., *Values and Peak Experiences* (New York, Viking Compass Book, 1970).

Maslow, A. H. *The Farther Reaches of Human Nature* (New York, Viking Compass Book, 1971).

Miller, R. S. & Eds. of New Age Journal, *As Above So Below* (Los Angeles, Jeremy P. Tarcher, Inc., 1992).

Redfield, James, *The Tenth Insight: Holding The Vision* (Warner Books, New York, 1996.)

Scholem, G. G., *Major Trends in Jewish Mysticism* (New York Schocken Books, 1941).

Scholem, G. G., *On The Mystical Shape of the Godhead* (New York, Schocken Books, 1991).

Spiegelman, Ph.D., & M. Miyuki, Ph.D., *Buddhism and Jungian Psychology* (Phoenix, Falcon Press, 1987).

Waters, F., *Book of the Hopi* (Harmondsworth, Penguin Books, 1984).

Wilber, Ken, *Sex, Ecology, Spirituality: The Spirit of Evolution* Vol. I (Shambala, Boston, 1995).

Wilber, Ken, *A Brief History of Everything* (Shambala, Boston, 1996).

Wilber, Ken, *The EYE of SPIRIT* (Shambala, Boston, 1997)

Williams-Heller, A., *Kabbalah: Your Path to Inner Freedom* (Wheaton, The Theosophical Publishing House, 1992).

Wolf, F. A., *The Eagle's Quest* (New York, Summit Books, 1991).

Woodman, Marion and Dickson, Elinor, *Dancing in the Flames: The Dark Goddess in the Transformation of Consciousness* (Shambala, Boston, 1996).

The Many Flavors of Psychology

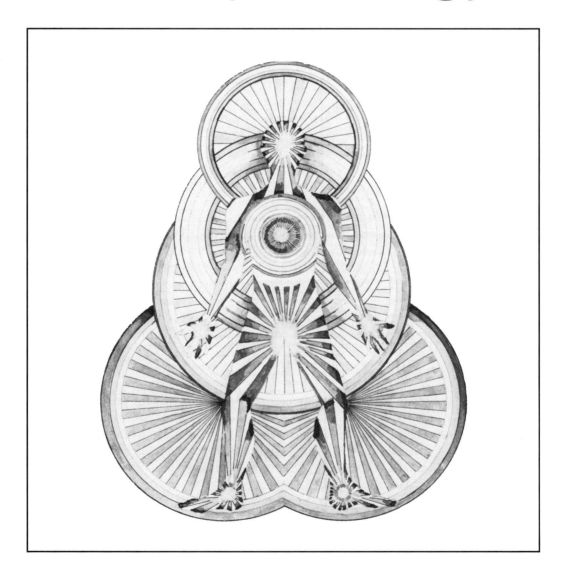

Section four

I now invite you to share ASELF's personal journey through the various levels and dimensions of Self. ASELF is a charming multi-faceted personality woven from the strands of personal and clinical experience. Various facets of her fictitious personality were carefully selected to illustrate the major psycho-social developmental tasks awaiting every individual. ASELF's story is one possible version of ALIFE. Yours is another. The underlying challenges and themes are universal, although the choices vary from individual to individual. ASELF's story demonstrates psychological principles and techniques, and offers a window into the psychotherapeutic process.

Once Upon ASELF

Once upon a time, many thousands of years ago, all was astir. The animal, plant, and mineral kingdoms heard that a special child, a human being, was soon to appear upon earth. And so it happened: ASELF was born to Mother Earth and Father Sky. Nature was ecstatic and knew great joy and celebration. From the very beginning, ASELF knew itself as two—an inner Dream Self and an outwardly directed Waking Self wrapped in the warm bosom of Mother Earth. And so it happened; *Puer natus est nobis*, ASELF had finally arrived!

Life With The Goddess

ASELF began her adventure in space-time under the protection of the Great Mother or Goddess. In the realm of the Great Mother, all that is needed is provided and given to the vulnerable, totally dependent human child, as long as she follows the rules and dictates of Great Mother's household. ASELF was born into a family of bounty and abundance, and never knew poverty or hunger. Her parents were caring and eager to please their newest offspring. A positive symbiosis and attachment developed between ASELF and Mother.* Mother was

---

*Persons of either gender can serve as caretaker, provider, and nurturer. Mother is therefore representative of any number of individuals who raise the young infant/child.

attuned and responsive to the needs of ASELF and provided a secure, trusting "holding environment" for ASELF to unfold. ASELF loved it whenever Mother sang to her and whispered softly into her ears. An invisible, strong, and loving bond was forged between the two for all eternity.

ASELF discovered that she was part of a large, extended family of parents, grandparents, sisters, aunts, uncles, and cousins. Her home was in the midst of luxuriant gardens and foliage. Sounds of chirping birds, barking dogs, splashing ducks, and flowing water filled her daily world. ASELF belonged first to her family and then to a larger community, and they also belonged to her. She felt very important and very special. And she was.

For a while, it seemed that Mother was an extension of ASELF, always responding, nurturing, and protecting. As ASELF grew, she realized that some of her wishes were not granted, and that certain activities she yearned for were displeasing or alarming to Mother. This was her first awareness that Mother and ASELF might not be the same person. It was a dilemma, as well as the first indication of a separate identity and a growing ambivalence to Mother. Occasionally, a frustrated ASELF had to acquiesce to Mother's needs and ideas, instead of her own. Mother would say, "No" and "Don't" whenever she feared that ASELF might get hurt. Otherwise, Mother allowed ASELF to discover her abilities and to try new skills. In this manner, her good-enough* mother allowed ASELF to explore and express her needs, while also protecting the young child from harm. Mother's trust enabled ASELF to develop trust as well, and to strive for autonomy. If a mother is threatened and overwhelmed by her child's needs, she might short-circuit the infant's initiative and drive toward autonomy by transferring her own distrust and anxiety onto the child. ASELF was fortunate; her mother had a strong sense of herself and so assisted her child in successfully navigating these crucial developmental stages. ASELF was quickly learning the rules and the ways of her parents' world.

At first, ASELF felt the anxiety of separation. She would only take a few steps away from Mother before re-approaching her side. Mother's presence made her feel secure and confident. Each time, ASELF was able to venture a few feet farther. She also discovered that her teddy bear and her security blanket were comforting "totems" whenever Mother was away. These transitional objects allowed ASELF

---

*"good enough" acknowledges there are no perfect parents or caretakers. What we hope for is one that is "good enough" and adequate.

to separate more easily from Mother for longer and longer periods of time. They also helped to consolidate her own sense of identity.

By the time she was two and a half, she felt quite independent. She was proud of her growing linguistic abilities. She readily used language to verbally communicate her needs and wants. She was learning how to dress herself, although the buttons refused to cooperate. And ASELF was enrolled in nursery school, where she discovered a whole new world of friends, rules, and opportunities.

ASELF was also the proud owner of an id, ego, and superego. These three psychic elements often seemed in conflict with one another, and ASELF was experiencing difficulty balancing the different perceptions and needs of each. Her impulsive, impatient "id" was eager to instantly gratify her needs. Her "superego" was quick to restrain her impulses. How the superego reminded her of Mother. They seemed to think alike, worry alike, and to express the same shouldn'ts, couldn'ts, mustn'ts, and don'ts.

ASELF's ego had not yet fully developed and was still learning how to achieve peace between the id and the superego. It was not an easy task. Gradually, the ego became stronger, and a separate, socialized Self emerged. Ego Self was the mediator between ASELF's outer and inner worlds. At first the two worlds were indistinguishable, just as ASELF and Mother had once been. As ASELF developed her own worldly personality, she seemed to lose a sense of her inner reality. She seemed more and more at home in the world of physical reality.

Gradually, ASELF forged an outer personality (persona) that was able to satisfy most of the expectations and dictates of her parents, her teachers, society, and even her peers. The cost of socialization and ego development came with a price. These processes required ASELF to relinquish some parts of her true personhood. ASELF had learned to please just about everybody, except herself. These undesirable, rejected, and split-off portions of ASELF had caused great discomfort to everyone, especially herself. These were the shouldn'ts, mustn'ts, don'ts, as well as the special, unique ways ASELF wanted to express her personality. Somehow, these aspects didn't fit in, and the only way that ASELF knew to keep peace was to deny their existence and reality. It seemed better to let these parts fade from view. Gradually... imperceptibly... these various discarded, personal Selves were submerged and repressed. Collectively, these newly disowned portions of ASELF formed a dark, unconscious aspect of her psyche called the "shadow." Although she was not consciously aware of what had transpired, ASELF felt a strange, puzzling sense of loss.

A multi-faceted ASELF was forming. ASELF now encompassed several different inner Selves, including Dream Self, Waking Self,

Ego Self, psychologist, persona, and Shadow Self.* Occasionally, ASELF would dimly remember a certain yearning or dream that seemed to disappear without a trace. Sometimes she wondered what had become of it.

ASELF didn't understand that all of her unique and special perceptions were still there, waiting in the recesses of her unconscious. Nor did she realize that these disowned aspects and sub-personalities could take on a life of their own and act without her consent during times of stress, crisis, and change. These hidden facets and voices would wait until ASELF grew and matured. They were the little treasures, the little pebbles that one found along the road and stuffed into numerous pockets for later exploration. Perhaps, one day, ASELF would rediscover them and let them transform her world.

The Patriarch Emerges

ASELF was maturing rapidly. She had successfully accomplished many outer world tasks, including socialization, toilet training, discovery of gender identity, feeding and dressing herself, and the development of language skills. Yet these tasks were not the only ones required of her. ASELF was also growing and developing psychologically, and how she fared in the inner realm would reverberate throughout her life.

As an infant, ASELF remained passive, receptive to Mother's all embracing, engulfing authority. This pre-conscious, pre-egoic state of unity is known as Matriarchal consciousness. However, between ages three and six, ASELF's psyche would undergo a crucial transition from matriarchal to Masculine (Patriarchal) consciousness, known as the Oedipal Crisis. She would need to relinquish her primal identity with Mother in order to define her own individuality. Only upon successful resolution of this struggle could ASELF hope to attain full Ego-consciousness.

Although ASELF was not consciously aware of the awesome psychic struggle she had recently completed (the Oedipal Complex), those around her were delighted to watch a bright, cheerful, capable, and confident little girl blossom and grow. No longer was ASELF just

---

*Dream Self visits the Dream world whenever ASELF is asleep, Waking Self is the awake, alert ASELF who can interact with the environment and respond to others, Ego Self mediates the needs of the impulsive id and the authority of the repressive superego, persona is the socially accepted aspect, and Shadow Self is the personal unconscious which contains an ever growing number of inner Selves, split off and disowned portions of her personality.

a physical being. She had moved from the pre-personal into the personal mode, and was now a psychological "I," fully aware of her "Self." She conveyed a special presence and poise that made her seem older and wiser than seven years.

A Separate Self Emerges—School Years

ASELF realized she was not a beauty. Frequently, she watched her younger, prettier sisters receive praise and attention for their appearance, while she was largely ignored. But ASELF felt a sense of relief when she realized that she was bright. This would become her path and the way she would create "ASELF." It would offer her freedom, independence, accomplishment, and even recognition. Yet, at seven, she had only the dimmest awareness of a future.

ASELF attended elementary and junior high school close to home. She was a quick, eager student whom teachers adored and encouraged. In emphasizing her rational, intellectual, and cognitive abilities, she had made a wise choice. How she loved to play with ideas and concepts, and follow wherever these thoughts would carry her. Her patient and willing father listened, hour upon hour, as she developed her thoughts and understanding on myriad subjects. Father would serve as her first mentor and guide. He recognized her brilliance, as well as her mystical, poetic side that she seldom acknowledged. He understood that ASELF would someday need to draw upon these other attributes and gifts in order to balance her intense intellectual pursuits. For now, he was content to encourage her curiosity and inquisitiveness.

At school, ASELF was well liked by her fellow students. Her peers voted her "the most likely to succeed." Intellectual recognition and accomplishment enhanced her sense of confidence and purpose. She was not particularly athletic, and participated in sports only when required. She clearly preferred to exercise and train her intellect. Yet her intensity and brilliance often caused difficulty between ASELF and her sisters. Her intellect became an impenetrable fortress that stood between her and almost everybody else in her family. Although ASELF desired closeness, her sisters wished distance and were unwilling to approach her. Because of her brilliance, Father held her in awe, while Mother and her sisters were jealous. ASELF did not understand the ambivalence she experienced in her own family. Mother favored her prettier sisters who willingly followed the more traditional path of feminine expression: passivity and obedience. Mother was often perplexed when it came to dealing with ASELF. She did not know how to approach her daughter, since their worlds and interests were so different. Father and ASELF were strangely on their own. She was grateful for Father's company and warm encouragement. Sometimes she worried what she would do if he ever withdrew his love and support.

It frightened ASELF to be emotionally dependent on others, and so ASELF would quickly push these fears and needs out of her mind. All of her disowned, often unacknowledged fears, anxieties, and worries were unconsciously banished to the shadow aspect of ASELF's psyche. She was gradually developing a rich repertoire of repressed inner children, sub-personalities, and disowned Selves which formed her personal unconscious. Inwardly and outwardly, ASELF was developing her multi-faceted personhood.

Adolescence

ASELF, at 12, was an adolescent and her hormones were exploding into awareness. Her body and her moods were rapidly changing. Some days she was amused and pleased with the new shape and curves of her body. Other times, she yearned to become the young, carefree girl she had always been. She couldn't quite make up her mind who she was. She had thought of herself as an "Ugly Duckling," and now she was becoming a Swan. She discovered that boys were no longer to be avoided. ... Boys were to be charmed and dated.

ASELF suddenly understood that her intelligence might limit her social success. Boys did not date girls who were smarter than they were. For the first time, ASELF diminished her intellectual abilities in order to gain social approval, and embraced the traditional feminine path. Her interests in scholastic activities began to wane and with them, it seemed, her aspirations for college and an academic career. She contentedly basked in her unfolding sexuality and the newly found attention it brought. ASELF had shifted the center of focus from her mind to her body.

High School

Now 15, ASELF had learned to deal somewhat rationally with her hormones and budding sexuality. During her last year in junior high school, she resumed her intellectual interests and again considered an academic career. She was relieved to discover her mind was still quick to grasp the rational and the logical, and she excelled in thoughts and subjects that often eluded her classmates. She realized how much she had missed playing with ideas. But she still experienced great anxiety and ambivalence over revealing her brightness to boys. She feared they would desert her, and she envisioned a life alone. She shuddered at the idea. Once upon a time, she had valued her intellectual gifts;

now she feared them. Father suggested that she search for a man who would value her intelligence and curiosity, as he did. He encouraged her to risk becoming more herself. Although Mother appeared worried, she said nothing to ASELF. Gradually, ASELF allowed her abilities to emerge and her grades to improve. She even passed the challenging entrance exam to the Bronx High School of Science. For now, she had succeeded in integrating mind and body, as well as social and academic parts of herself.

ASELF enjoyed Bronx Science and felt a firm and comfortable fit with the rest of the students. She made a few friends and had enough dates to fill her weekends. Nothing serious; she was only 16. She was merely practicing her interpersonal skills for a special someone, some day. ...

She particularly liked chemistry, yet vehemently disliked physics. Mathematics was occasionally troublesome, but manageable. Her first two years at Bronx Science flew by quickly. Although ASELF approached her senior year with confidence, she noticed a slight tinge of anxiety whenever she thought about college. Which college would she attend? What did she want to study? Decisions, decisions, decisions.

She began to recognize two persistent inner voices. One, the *critic*, enjoyed reminding ASELF that she was "inadequate and simply not good enough for a career in science. Who did she think she was anyway?" This voice knew more than a dozen ways to tell ASELF her limitations and faults, and had nothing positive or nice to say to her. The other, her *saboteur*, was quite adept at getting her to waste time, so that there was less and less time for studying, reports, and filling out applications. ASELF was scared. Nothing like this had ever happened to her before. She began to dread the very subject of college and her future. Whenever she avoided these issues, she felt a sense of control and relief, which unfortunately served to reinforce her decision not to think about college for a while. Again, her grades began to drop and, one day, the school counselor asked ASELF to come and visit with him.

ASELF visited with the school counselor several times over the next four weeks. He realized that ASELF was diminishing her opportunity to go to college... any college. ASELF was scared, plain and simple. She was not ready to deal with separating and individuating from her family. College meant living away from them and a need for greater independence. ASELF was struggling with dependency needs versus autonomy. The level of separation anxiety was very intense at times, as was her fear of success.

ASELF acknowledged another conflict raging deep inside, one between her academic needs and her social needs. She was caught in either/or thinking. The fear of social rejection due to her academic success was familiar to her. She had dealt with it during junior high school and thought she was finished with such concerns. She was learning that her issues were "her issues for life" and that her challenges would recur in many different forms and at various stages in her life.

Her surprised parents were invited to share a session with the school counselor and ASELF. They willingly reassured and assisted her as she prepared for her entrance exams. Once more, with Father's encouragement, she took a risk and allowed herself to concentrate and to achieve. Her grades went back up, the voices and actions of the critic and the saboteur quieted, and everybody relaxed.

Meanwhile, ASELF learned some relaxation techniques and enrolled in a preparatory class for the college boards. She began to honor and allow her negative thoughts, and she started to acknowledge her fears rather than avoid them. And she learned how to reframe her fears and worries into more constructive, realistic thoughts about herself. She again dealt with her fear of being smarter than her dates, and realized that she was giving these young men too much power over her aspirations. She took her exams and learned a few weeks later that she had scored well. She felt a warm, calm sense of relief flow through her. She began to smile. ASELF passed her qualifying exams and she walked into her future, feeling strong and elated.

College

ASELF was accepted by several colleges and universities. She decided to go to U.C.L.A. even though it was far away from her family in New York. Characteristically, she chose to confront her separation and individuation issues directly.

ASELF loved new beginnings. She fully immersed herself in the psychobiology program at U.C.L.A. She was so busy, she really didn't have time to miss her family. The cognitive aspects of her mind were very much in the forefront, helping her to achieve and excel scholastically.

At Christmas, ASELF returned home, very satisfied. She relished the warmth of her family and the activities of the holidays. And she looked forward to returning to the west coast. For now, she had struck an important balance. In just three short months, ASELF had created a new life and world in Los Angeles.

Her college years flew by. In her senior year, ASELF had already passed the necessary graduate entrance exams and had sent applications to several doctoral programs. She preferred to stay at U.C.L.A. and was overjoyed when they accepted her into the Department of Neuroscience. She had made many contacts and friends among the U.C.L.A. faculty and felt that to remain in Los Angeles would be advantageous to her career. She completed all the requirements for the bachelor's degree in psychobiology. Following graduation, ASELF vacationed in Europe for a month before returning home to visit with her family.

Two of her sisters were now married. ASELF sensed her parents' concern about her future, as they expressed their hopes she would one day marry and start a family of her own. They had never before openly addressed such issues, and so ASELF was startled by their worry. Until this moment, she had been content with the rhythms and thrust of her life. Suddenly, the familiar conflict between social and academic needs reappeared. This time, it was Mother who raised these issues with ASELF, and Father who passively remained silent. She felt the sting of her parents' disapproval. A worried and saddened ASELF returned to Los Angeles. She felt that Father had abandoned her by his silence. In her despair, a disquieting feeling of inadequacy began to take hold. The thoughts and attitudes of her critic and her saboteur sub-personalities flourished in this environment of insecurity. Indeed, the two inner voices took on a life of their own and proceeded to torment her over the next few weeks. ASELF found it difficult to sleep and to concentrate. She questioned her abilities and considered dropping out of the graduate program even before classes began. She was profoundly depressed.

For several weeks, ASELF continued to feel angry and confused. School resumed, and she was rapidly falling behind in her studies. There was something very familiar about these issues, and she sensed that separation and individuation from her family was at stake. ASELF made the decision that she, not her parents, would choose her path. Courageously, she sought assistance in unraveling the fears that engulfed her. Her "psychologist" inner voice encouraged her to hear her parents' concerns rather than to angrily dismiss them. She no longer denied her own needs and desire for companionship. Gradually, ASELF initiated contact with her friends, attended concerts, joined a gym, and looked for ways to balance her many needs. She resumed her life and recognized that an *either/or* choice was neither healthy nor workable. Now she would balance the intensity of her academic pursuits with social activities. With a great surge of relief and understanding, ASELF felt ten feet tall.

A Family of Inner Selves

ASELF was learning to recognize her projections and behavioral patterns, and give expression to her attitudes and feelings about important issues and events. Over time she uncovered an entire community of inner Selves and recognized her multi-faceted nature. She was learning to actively dialogue with numerous sub-personalities, inner Selves, and paradoxical voices dwelling within the hidden domains of her personal unconscious. All were parts of her being. She courageously allowed her Career Self to develop, and joyously released many previously disowned social aspects into her life.

In the outer, physical world ASELF was sister, daughter, student, friend, room-mate, and lab-assistant. Her personal unconscious (shadow) similarly contained multitudinous fragments and patterns of energy. ASELF gradually characterized these inner voices as critic, saboteur, researcher, executive, the aggressive one, defiant child, mystic, funny-lady, and sexy-woman. A Mother Self revealed her wish to one day nurture ASELF's future physical children, and promised to be available whenever she was called upon. Good, bad, indifferent, fearful, as well as possible or future Selves surfaced. ASELF sometimes thought her psyche contained a cast of a thousand characters. Many competing emotions pulled ASELF in opposing directions. It was difficult to listen to all of them at once. ASELF decided to grant the role of facilitating Self to one eminently qualified inner voice. This very wise and altruistic Self was able to successfully recognize, coordinate, and integrate the diverse, paradoxical attitudes and inner Selves residing within. The myriad strands and colors of her multi-faceted personhood were woven into an exquisite tapestry of personality called "ASELF."

ASELF worried that she might uncover a more complete personality called an *alter* and be diagnosed with Multiple Personality Disorder (MPD).[1] She sought the advice of her psychologist. Annie quickly dispelled her fears. She explained that in a normal personality, our "I" and "me" are woven from a multiplicity of energy patterns and inner Selves. Each sub-personality was an incomplete fragment or facet of a more inclusive personality. But in MPD, the composite "I" becomes a composite "we," which incorporates as well as forms around a core of (more or less) complete, distinct personalities. Annie assured her that she didn't have MPD. ASELF took a deep breath, and was relieved to know she was okay.

Reassured, ASELF continued to explore her multi-faceted Selfhood. She was impressed with the richness and complexity of the psyche, and amazed that so many strands of consciousness went into the creation of a Self. The changing panorama within her mind reminded ASELF of a kaleidoscope. The many fragments and components of her personal unconscious seemed to blend and merge into a never-ending array of psychic patterns. Each one affected her perceptions, her moods, and her understanding of the world. ASELF recognized that she was both the One and the Many. The grandeur, enormity, and complexity of a multi-dimensional Selfhood did not escape ASELF. Something deep within the core of her being seemed to resonate with the truth that was revealing itself. An inner knowing began to stir and hum. ASELF had taken her first steps into a very special world and it was enough to simply acknowledge its existence. Someday, she would more fully awaken and explore its secrets. For now, she needed to return to the physical world of the senses.

Healing A Family

ASELF wished to heal her relationship with her family. She spoke to her father about her fear of his abandonment and he reassured her that he would always be there for her. Father *voiced* his satisfaction with ASELF's decision to pursue both her career and social needs. Imbued with Father's support and encouragement, warmth and calm once again flowed through her. She realized how important Father's acceptance was to her sense of well-being.

ASELF and Mother encountered greater difficulty in approaching one another. For a long time, ASELF had been aware of anger and disapproval from Mother, and for a long time, she was unwilling to deal with it. Gradually, Mother and daughter were able to speak their emotions and concerns to one another. ASELF was surprised to learn that Mother was actually jealous of her. Mother revealed that years earlier, her business Self had also wanted to venture into the world. She had never mentioned these hopes and longings to Father, for she was certain he would not approve. Like so many generations of women who preceded her, she had passively given up her dreams. She had projected her disappointment and anger onto Father, and later, displaced her fear and resentment onto ASELF.

Father was aware of Mother's frustration and anger, and sensed that she might have been happier if she had some outside interests of her own. Yet, he was reluctant to encourage Mother for fear of pressuring her to meet *his* needs, rather than *hers*. Instead, the two had made an unconscious agreement never to talk about her needs (or his) for autonomy and individuation. It was too scary for both of them to confront.[2]

When ASELF showed interest and ability in ideas and thoughts, Father quickly encouraged her. Mother felt angry and jealous. Why didn't he discuss such ideas with Mother, or encourage her pursuits? Each parent remained unaware of the needs and desires of the other, and each projected their disappointment and passivity upon the other. Father openly supported ASELF, as he had once hoped to encourage his wife, while Mother grew to resent her own daughter's tenacity and perserverance because it reminded her of unfulfilled dreams and longings. Mother was jealous of all the attention Father heaped upon ASELF. She wanted some attention too. It was safer to project her displaced anger upon ASELF, rather than openly express it to Father or to confront it in herself.

With ASELF's expression of her feelings and needs, and her courageous determination to talk things out with her parents, the entire family began to heal and grow. Father felt freer to support ASELF since he was no longer caught between his wife and his daughter. Father also began to encourage Mother to develop some of her own abilities. Mother delighted in her accomplishments and Father overflowed with pride. He even allowed some of his *own* dreams to emerge and become actualized. Gratefully, Mother and daughter were no longer opposing one another. ASELF's parents were uncovering their own inner voices and dreams and working through their own processes of separation and individuation. ASELF's confidence returned and she no longer went to see her psychologist. She and her entire multi-faceted community of inner and outer Selves confidently went on with their lives.

ASELF Meets Science

ASELF was fully immersed in her graduate studies. She expected to complete the requirements for a doctoral degree in Neuroscience by the end of the year. She delighted in the parallel processes of learning about her physical neurological system while uncovering her inner family of Selves. Both systems affected her perceptions and behavior in the physical world. It was the first time she was aware that events in the outer world might parallel inwardly occurring events.

ASELF found it astonishing that old paradigm Science failed to acknowledge the existence of "Mind." Unable to see or measure the mind with the tools and techniques available, science chose to simply deny its reality. ASELF realized that in this way, science had disavowed the spiritual and sacred roots beneath and beyond our universe. She understood that the connection between inner and outer worlds had been severed; humanity was no longer being nourished or replenished by its compassion and wisdom. By ignoring the innermost regions of Mind, she knew that science had ignored and denied the very essence and core of our humanity. Stunned and shaken, ASELF recognized that science had effectively divorced humankind from the sacred and the Numinous.

ASELF was well aware of science's views. She carefully protected her inner world from her scientific colleagues and peers, fearful that they would consider her superstitious, silly, and filled with magical thinking. She understood the necessity of separating her inner world from her outer reality, in order to be a "scientist." She wondered why science avoided subjective modes of knowing, choosing instead to remain impassionately objective. She found it impossible to remain so uninvolved. She wondered what might happen if inner and outer realms were allowed to interact with one another. Wouldn't life be richer and more satisfying? She thought so. Although ASELF found these ideas exciting and stimulating, the answers always lay just beyond her grasp and comprehension.

ASELF was awarded a prestigious grant. She was grateful and pleased. Two of her research papers had recently been published and she received invitations to lecture at several universities across the country. She eagerly accepted the opportunity, and felt exhilarated by the recognition given her work.

While lecturing at Harvard, ASELF met Rising Star, a fourth year medical student, who attended all of her presentations. On her last evening, she accepted his invitation to dinner and had a wonderful time. ASELF returned to Los Angeles, but she couldn't get him out of her mind. She found herself fantasizing and thinking about him constantly. A few days later, she received a telephone call from Rising Star. He had accepted a hospital rotation in Los Angeles and would be staying with friends for two months. Could he see her while he was there? She was breathless. She encouraged Rising Star to come and even volunteered to show him the sights and sounds of Los Angeles. She began to count the days until his arrival. ASELF was in love.

In Love

ASELF and Rising Star became inseparable. They enjoyed similar interests and activities, and delighted in encouraging one another's careers and visions. They created a world of their own, and then it was time for Rising Star to leave. ASELF dreaded the pain of separation from her beloved. She could hardly bear it, so she immersed herself in her work and decided to complete her dissertation as rapidly as possible. ASELF knew that she wished to share the rest of her life with Rising Star. But she worried he would forget her once he returned to Harvard. Rising Star found that he missed ASELF more than he had ever missed anything or anyone else. He, too, feared that he might be forgotten.

As days slowly went by, the two did everything possible to reassure one another of their love. They made plans to meet during the next holiday. This time, ASELF flew to Cambridge and stayed with Rising Star. She met his friends and explored his favorite haunts. They fell more deeply in love. Now, time flew swiftly by, and ASELF returned to Los Angeles, more lonely than ever.

ASELF had only to write her dissertation; Rising Star had medical boards to pass. Rising Star suggested that ASELF write her dissertation in Cambridge so that they could be together over the next months. She accepted, for she knew that they needed more time together in order to assure their future. Her professor supported her decision and she lost no time in making the necessary arrangements. ASELF arrived in Cambridge on a cold, rainy day. She wondered what would happen if their relationship didn't work out.

Everything went extremely well. Their time together, the daily sharing of events and activities was wonderful. Graduation from medical school was now two weeks away. Rising Star's parents and brothers were planning to attend the ceremonies; ASELF would finally meet them. She was already deeply invested in the relationship and suddenly felt even more vulnerable. *"What if they don't like me or approve of me?"* She spoke her fears to Rising Star, but he laughed. "Of course they'll love you, just as I do." ASELF worked to keep her feelings under control and tried not to worry. It turned out that Rising Star's family adored her, and that she liked them as well.

ASELF's mother and father were both amazed and delighted that their daughter was in love. After Rising Star's graduation, they invited the young couple to New York. Now it was Rising Star who worried that he would not be accepted. But ASELF's family accepted him immediately. Upon their return to Cambridge, Rising Star proposed to ASELF and she joyfully accepted.

Wedding Bells

During the next few months, ASELF planned her wedding and completed her graduate studies. And so it happened: ASELF married Rising Star.

They relocated to San Diego, where Rising Star opened a family practice and ASELF accepted a research position. They were happy. With great joy, they welcomed their children, Harmony and Blessing, born a year apart. And so it happened: They were a family.

Rising Star's practice continued to grow and it kept him very busy. Gradually, more of the family responsibilities were thrust upon ASELF. She did whatever she could to assist her husband's career. Yet in her eagerness to help him, she was gradually giving away the time she needed for her own career and interests. Rising Star seemed to spend less and less time with her and the children, and she sensed they were emotionally drifting apart. Whenever she voiced her fears, Rising Star would impatiently change the subject. ASELF slowly withdrew from the man she loved. She went to see Annie, her psychologist. ASELF felt that her marriage and a portion of herself were lost.

Chaos and Uncertainty—A Marriage Ends

ASELF was ashamed that her marriage was failing, and too readily assumed all the blame for its failure. In so doing, she shielded Rising Star from any responsibility. ASELF had willingly pleased Rising Star, as well as eased his way, yet asked for little in return. Like her mother and generations of women before them, ASELF failed to ask for what she needed or wanted. Unconsciously, she had re-created her parents' marriage and realized her worst fears. She began to understand she had chosen a husband who was unwilling to meet her emotional needs. ASELF had assumed Rising Star, like her father, would guide and encourage her intellectual pursuits and career, and was dismayed when he did not.

Unresolved patterns and dramas originating in Rising Star's family were also unconsciously re-enacted and imposed on ASELF. His mother had sacrificed her career to help his father's, and Rising Star confidently expected ASELF to do as much for him. *"After all, isn't that what marriage is about?"* he asked. When ASELF refused to give up her own career after the children were born, Rising Star became angry and distant. He felt betrayed.

With great insight and pain, ASELF acknowledged that she had been reluctant to support and encourage her *own* career. The familiar tasks of separating and individuating from those she loved was again challenging her to acknowledge and develop her abilities and gifts. A deeper, more *authentic* ASELF struggled to emerge. Weary but determined, she reclaimed her path and continued her journey.

The Work of Psychotherapy

ASELF worked hard in psychotherapy, and gained insight into her patterns and her choices. She eventually accepted responsibility for them. With her psychologist's support and assistance, ASELF resolved the present crisis. She realized that *intimacy* is really *honesty*. She let go of blame, judgment, and self-righteous indignation, and courageously dealt with the inevitable.

ASELF and Rising Star were unable to resolve their differences, no matter how hard they tried. Sadly, they admitted that their marriage was over. Rising Star moved out of their home and a great emptiness moved in. ASELF began to pick up the pieces of her life. She had two children, a career, and a future requiring her attention.

Perennial Philosophy

Her psychologist, Annie, recognized that ASELF needed a more expanded framework in which to grow. She introduced ASELF to the world of Transpersonal psychology, which encompassed the totality of human experience. Sensing that ASELF was moving from the personal to the shared collective levels of personhood, Annie told ASELF about infinite realms of awareness and non-ordinary states of consciousness, along with their potential for healing, transformation, and spiritual awakening. ASELF learned that spiritual traditions of every age and of every people embrace a shared core of ideas, known as the Perennial Philosophy (See Table 4). The Spectrum of Consciousness, a psychological version of the Perennial Philosophy, is the heart and soul of all Transpersonal psychologies.[3] ASELF felt herself being powerfully and seductively drawn to the mysteries of life. She intuitively sensed their deeper meaning and purpose. Excitement and anticipation begin to stir within her. The mysterious and the Numinous were touching her Soul.

Annie shared many provocative ideas and concepts with ASELF. She told her, "Consciousness is the creative principle that forms and informs our physical universe." ASELF was perplexed and pondered this idea for weeks. Eventually, ASELF decided that it meant, "Consciousness is primary," since it precedes as well as creates all matter. Annie told her, "Consciousness seeks to know *Itself* through form and experience."

Table 4

PERENNIAL PHILOSOPHY is a core of enduring and universal spiritual beliefs. As ancient seers and mystics explored hidden, invisible dimensions while in non-ordinary states of consciousness, they uncovered these noble Truths. The fundamental concepts found in every major religion and spiritual tradition are:

1. A transcendental unity as Source and Sustainer of all Existence. Consciousness, not matter, is primary.
2. Creator is not separate from creation. Consciousness is a WHOLE, a ground or field in which all existence is immersed. This whole is also known as Universe. The act of creation is the cause of apparent duality between Observer and Observed. We know Universe and SOURCE not from a distance, but from BEING in it.
3. Consciousness assumes a multiplicity of forms and explores infinite realities. Consciousness is endlessly creating, emerging, and evolving.
4. All describe a non-linear spectrum (holoarchy) of infinite levels, in which higher levels form, sustain, and enfold lower ones. Essentially, all levels are interconnected and interpenetrated. Each level is endowed with its own grade of consciousness, intelligence, and meaning.
5. Life is a journey of self-discovery, fulfillment of earthly challenges, and spiritual return to the ONE WE ARE.

References: Friedman, "Bridging Science and Spirit."
Huxley, "The perennial philosophy."
Walsh and Vaughn, "Paths Beyond Ego."

ASELF was beginning to understand that she was a part of a whole called Universe. She began to think in wholes rather than in parts, and she noted many parallels among ideas of psychology, new paradigm science, and spiritual traditions. Despite the varied metaphors and language used by each discipline to express its ideas and knowledge, ASELF was certain that many of the ideas and concepts were interchangeable. She had discovered a fundamental oneness behind all the fragmentation and separation of the world. A subtle, rather profound inner shift in her perception of Self and reality was gradually taking place.

Annie identified the very essence and core of her being as "SELF." She informed her, "A fundamental split is occurring, allowing one part of the SELF to observe the other, who acts." ASELF realized that such a division permits only one portion of the SELF to be known, since the *Seer* always remains unseen. She thought that humankind's relationship to its Source also reflected this fundamental split. Annie smiled and remarked, "All divisions of the SELF are arbitrary, for All are parts of the whole." Although it appears that two parts of SELF resulted from this split, there was still only one. Annie used a mirror to illustrate how it reflects one object as two.

Annie further stressed that the full essence and flavor of any spiritual experience can never be conveyed in words. It can only be felt and experienced. She said the true meaning of *Gnosis* always occurs on the other side of words. Like the seer, the mysterious Source of all existence remains forever incomprehensible and unknowable.

Annie shared another propitious idea, "SELF is God dwelling within us."

She told ASELF that the spiritual journey, in its many forms, encourages us to heal the split(s) within our psyche so that observer and actor, again, become one.

Wide-eyed and curious, ASELF whispered, "Who is God?"

Annie said that no one could really answer that question. Then she said: "Awareness and understanding of God changes and evolves day by day, and even moment by moment." Suddenly, ASELF had an amazing insight: "To know SELF is to know God." ASELF allowed this awesome insight to flow through her body and mind and she felt herself tremble.

ASELF resonated with these profound ideas. But she was also aware that such views radically differed from Newtonian-Cartesian models of the psyche. She knew that such paradigms insisted, "Consciousness is merely a consequence of neurophysiological processes occurring in the brain and nervous system." Since matter is deemed primary, most scientists claim consciousness is essentially limited to matter.

ASELF told Annie that traditional psychologies generally view non-ordinary states of awareness with suspicion, deeming them pathological and dangerous. As a result, the healing and curative potentials of altered states of consciousness are denied. ASELF angrily declared that too often, "Unsuspecting psychiatrists sedate, medicate, and suffocate the awakening spirit, and silence the wisdom it offers."

ASELF's own research centered around regressive and unstable pathological mental diseases and she acknowledged that psychotic states did, indeed, exist. But she also understood that an either/or dichotomy with regard to altered states of consciousness was incorrect. She recognized the value and need for preparation and guidance when exploring inner realms of consciousness. But ASELF intuited that there was much to learn from such non-ordinary states. Her personal experiences, and those reported by others, revealed amazing potential for change and growth.

ASELF had entered the world of Transpersonal psychology. The spaces and realms described were strangely familiar and she felt comfortable with these ideas. She particularly liked Jung's theory of individuation, because it recognized the natural and spontaneous urge of Psyche toward wholeness. ASELF instinctively felt that the process of individuation was both healthy and necessary for psychological well-being and self-realization. She made a decision to trust the wisdom of her unfolding inner Self. ASELF had experienced the *at-one-ment of individuation*; she felt *at-one* with herself, *at-one* with humanity, and *at-one* with Universe.

The Archetypes

One morning, ASELF awakened from a vivid dream in which a young child kept beckoning to her. She was deeply impressed by the dream and recorded it in her journal. Annie thought the young child was the *Puer*, or inner child *archetype*, symbolizing healing and wholeness, beginnings and endings, and signifying the potential reconciliation of psychic elements. Annie was especially pleased with this primordial dream image, as it conveyed a promise of ASELF's future growth and transformation.

ASELF was having difficulty distinguishing the archetypal inner child from the many inner children and sub-personalities that formed her own multi-faceted personality. Annie told her, "Sub-personalities and disowned, repressed elements of the *personal unconscious* reflect an individual's history, experiences, memories, desires, and dreams. These are the inner children who dwell in the 'shadow' and participate in the events and dramas of the everyday world. Early phases of psychotherapy take place upon this personal level, and strive to heal the split between ego (persona) and shadow." Annie reminded ASELF that the spiritual path is arduous and strenuous and requires a strong, healthy ego. She also cautioned that preparation and guidance are necessary for a safe journey to the deeper levels of Psyche.

Annie revealed that the archetypal symbol of the Puer emerges from deeper levels of the *collective unconscious*.[4] These inherited, primordial forms, belonging to all humankind, powerfully influence and pattern our perceptions and behaviors. Since they are shared by all humanity, the archetypes are collective and transpersonal in nature.

ASELF noted a correspondence between the genetic blueprint encoded in the DNA molecule and the archetypal forms of the collective unconscious. The former guides physical growth and development, whereas the latter guides evolution of the psyche. Both offer intricate possibilities for personal development, and both interact with the surrounding environment in which a person lives. ASELF began to suspect there are no accidents in the universe.

Annie described two important contrasexual archetypes: the anima and animus. Jung discovered that the feminine *anima* resides in the male psyche, while the corresponding masculine *animus* dwells in the female psyche. Annie agreed with Jung that Psyche is, therefore, androgynous. Annie believed that harmonious reconciliation of the masculine and the feminine cannot occur until each is completely differentiated. The masculine separates and develops during the rule of the Patriarch. Once this has occurred, the stage is set for the emergence of the Feminine. As Self matures and Psyche evolves, masculine and feminine aspects yearn to approach one another so that they may unite. *The two are now one and "we" become whole.* According to Annie, this is the meaning of the *Sacred Marriage* between the masculine and feminine portions of the psyche. It is a gateway to the sacred.

ASELF realized that instead of completing these psychic tasks, people project their archetypal anima or animus upon lovers and spouses (children too), insisting they be whomever we need them to be. When they fail, we get angry and judgmental. We heap blame and guilt upon them, and self-righteously proclaim, "It is you who is wrong, not me!" We do not recognize that it is our own unconscious processes that need attention. ASELF thought, "So many of us wait until our relationships fail, and love turns to hate before asking, 'Why?' and 'What went wrong?'" And suddenly, she felt very, very sad.

Rising Star and ASELF had projected and re-enacted their family patterns and dramas throughout their marriage. ASELF wanted a husband who would support and encourage her career, as did Father, whereas Rising Star expected a wife to devote herself to satisfying his every professional need. No matter how hard they had tried, they were unable to change one another. Sadly, each came to the realization that the person they wanted was not the person they had married. Their paths were moving in different directions and each needed to *let go* of the other in order to survive and grow. ASELF's new insight helped her release anger and blame, and her heart filled with renewed peace and compassion.

ASELF was moving out of the confusion and chaos of divorce. She was beginning to heal and so were her children. Rising Star had re-married shortly after the divorce was final, and he moved to another state. ASELF and the children missed him—a lot!

ASELF's life was frenetic as she sought to balance single parenthood, career needs, and everyday chores. She was always exhausted. Her life assumed its own rhythm and the days flew by. Two years had passed since her divorce. To her amazement, she had survived.

Annie had been a tremendous help in supporting her through the trauma and pain of divorce. She had opened new vistas and introduced new concepts and ideas. With Annie's assistance, ASELF had discovered her multi-faceted personhood and initiated the healing and unification of ego and shadow. ASELF was moving beyond the issues of the id, ego, and superego, and toward the challenges and needs of her Existential Self.

Existential Self still deals with personal issues, but does so with enhanced awareness and is fully cognizant that she is part of a greater Selfhood. Consciously aware of her projections, needs, and feelings, and able to take responsibility for her actions, she has moved beyond dependency, blame, and guilt. Existential Self desires freedom of choice and autonomy, and acts from a space of authenticity and integrity.

But Existential Selfhood cannot be attained unless Ego Self agrees to relinquish its present position of centrality, as well as to subordinate its interests in deference to the needs of a greater SELF. By shifting its position, Ego Self is no longer alone and alienated. It has become part of a whole. Initially, "Ego" resists any attempts to release its tenacious grip on Psyche, for it fears it will be destroyed in the process. Despite the ensuing internal struggle, Ego Self is never abandoned or destroyed. Considerable time and effort were necessary before ASELF was ready to complete this arduous psychic task. And then one day, she suddenly let go. And so it happened; Existential ASELF had arrived.

ASELF moved into this expanded personhood. She discovered that her new insights and achievements followed her into the world of physicality and enriched her days and activities. Both inwardly and outwardly, ASELF was being challenged by issues of mortality. Many times during the next year, she was to confront her vulnerabilities, limitations, and her need to be "rescued."

Mom was in a serious car accident and died suddenly. Father and ASELF were overcome with grief. ASELF flew home to support Father during the first few weeks of mourning. Mom's untimely transition raised many questions about death, personal meaning, and the purpose of life. ASELF was questioning the very nature of reality and wondering, *Who am I? Why am I here?* and *Why?*

ASELF experienced a great sense of loss and emptiness following her mother's death. She painfully acknowledged a loss of fulfillment and vision in her own life. And she was aware that she was still mourning the loss of her marriage. Ever since the divorce, ASELF had put her life on hold, and simply *endured*.

It seemed that by her death, Mom was urging ASELF to live more fully, and to move beyond Ego and Existential Selves. She challenged her daughter to create a future Self that would again know joy and fullness. The familiar issues of separation and individuation were again challenging ASELF to heal, to grow, and to transcend her present sorrow.

Father moved to San Diego to be near ASELF and his grandchildren. It was good for all of them. His presence and love warmed ASELF's heart and lessened the emptiness inside. Her research findings were widely published and she was gaining recognition in her field. She proudly accepted the position of Dean of Neuroscience, and hoped that it would give her an opportunity to shape academic and scientific policy.

Several months after her mother's death, ASELF had a dream about Mom. In this dream, Mom told her that she was well and happy. She implored ASELF to go on with her own life and to continue her exploration of the unknown regions of Mind. ASELF wondered where dreams originate. How could Mom still communicate with her? Did consciousness survive, somehow? Was it really Mom or another part of her unconscious telling her to grow? ASELF was overwhelmed with questions.

Stirrings of the Feminine

Once more, she sought her psychologist's guidance. Annie was very pleased with ASELF's inquiries, for they proclaimed a new stage of psychic evolution. An enlightened and mystical aspect of ASELF's psyche, dwelling in the collective unconscious, began to stir. This portion of ASELF was inviting her to seek community, to acknowledge her interdependence and interrelatedness with all forms of consciousness, and to transform Separate Self into Relational Self.

Annie revealed that Transpersonal Self whispers to us during dreams and visions, non-ordinary states of consciousness, and moments of intense creativity. This archetypal governing principle awakens whenever we open our hearts and feel compassion toward another. Emergence of Transpersonal Self signals a shift from archetypal Patriarchal toward feminine aspects of the psyche. Whereas the Patriarch promotes the personal, separate, and the individual, the archetypal Feminine embraces altruism, interdependence, and relationship. The emerging Feminine is a hallmark of an evolving psyche. Transpersonal Selfhood introduces us to the Feminine, brings us wholeness, and awakens us to SPIRIT.

ASELF found a spiritual teacher, Mai, who taught her the art of meditation and contemplation. She learned that these techniques are pathways to self-realization and the transpersonal realms. With Mai's help, ASELF began a daily practice of meditation. Mai taught ASELF to follow her breath as it moved in and out, and to honor her thoughts and feelings without judgment. ASELF learned to sit quietly and to just "Be." Gradually, she attained a state of mindfulness and clarity. Inner silence and tranquility gently flowed through her psyche. She no longer felt bound by time and space. She knew herself as the awareness before any thought and sensation, and beneath all existence. And so it happened; Transpersonal Self emerged.

The emergence of Transpersonal Self created a different relationship between reality and ASELF. She noticed that an inwardly calm ASELF acted with calm in the physical world. Irritability inside always manifested as an irritation "out there." If she felt angry, her mind searched for someone or something in the *real* world to be angry about. The phrase, "As above, so below," aptly described the way her inner and outer worlds influenced and reflected one another. ASELF was also aware of her many inner Selves, their attitudes, feelings, needs, and projections. She responded more and reacted less. She experienced a oneness with Universe and a renewed sense of belonging and purpose.

ASELF had evolved from pre-personal consciousness and dependency upon the Great Mother Goddess of early childhood into an independent, multi-faceted personal consciousness ruled by the Patriarch. With the emergence of the Feminine, ASELF realized her interdependency and interrelatedness with all of creation, and expanded into the realms of Transpersonal consciousness. ASELF, seeking the Numinous, yearned to become The One We Are.

A more peaceful, gentle ASELF was unfolding. She had an inner radiance and certainty about her, and seemed more at ease with herself. Everyone noticed her transformation.

ASELF embraced global issues and recognized the urgent need to heal our planet. She knew it was essential to live in peace and harmony with Mother Earth, and that time was growing precipitously short. She was actively engaged in healing and serving the planet. ASELF hoped to learn from the world's indigenous tribes as well as from her fellow scientists. Sometimes ASELF wondered if there was such a thing as a transpersonal scientist. She thought she would like to become one.

ASELF was asked to attend a very special conference, along with representatives from many disciplines and from many peoples. Scientists, parlimentarians, educators, social scientists, doctors, spiritual leaders, and indigenous tribesmen voiced their fears and hopes for our world and our species. This distinguished forum of scholars wished to develop a new scientific discipline devoted to the study of consciousness. All considered a spiritual and sacred science essential to the continued survival of humankind. Excitement and possibility were everywhere.

ENOUGH, a respected parlimentarian from Canada, was at this conference. Recently widowed, and father to 14-year-old Ananda,*

---

*ANANDA means exquisite bliss and joy.

ENOUGH was a kind, sensitive, and fun-loving man about ten years older than ASELF. A mutual friend introduced them to one another at this conference. During the next year, the two met many times to share and develop a new science of consciousness. Slowly, almost imperceptibly, ASELF and ENOUGH fell in love.

ASELF was scared. She still felt the pain and loss of her first marriage, but she had transcended the anger. It was time for ASELF to let go of the past and move into the now.

ASELF's emerging Feminine was receptive to this loving relationship, which offered a wealth of opportunity for discovery and transformation. Transpersonal ASELF adored ENOUGH and equally embraced this new romance. It was Transpersonal ASELF who urged ASELF forward, and encouraged her evolution from Separate Self to Relational Self. ASELF inhaled a deep breath and decided to take the risk. She opened her heart and let it fill with love.

Months later, ENOUGH and ASELF gathered their children and families together. They announced their wedding plans. The children were now part of a larger, blended family. Father was especially happy, because he realized that his daughter had grown to accept herself as woman, mother, wife, and scholar. She had created a rich personality and had learned to balance her separate worlds. Father knew that she had found a man who would honor her and support her journey. And so it happened; ASELF and ENOUGH became husband and wife.

ASELF was awakening to the Numinous. She had grown into a fuller, more evolved Self and had journeyed through the Path of the Heart. She recognized the incredible multiplicity of the One and its expression in the Many. She embraced her uniqueness and one by one had claimed her many Selves. She recognized Waking Self, Dream Self, Id, Superego, and Ego Self, Existential Self and Transpersonal Self. She was also wife, mother, daughter, sister, scientist, and mystic. She had, indeed, become the One and the Many.

ASELF wondered if it was possible to evolve beyond Transpersonal Self, and what her world would be like if she did. She began to hear SPIRIT's soulful call; it was inviting her to awaken and to travel the Path of the Heart. Humble, loving, and compassionate, ASELF was moving beyond the world of space and time. She embraced the sacred mysteries of Universe.

Puer natur est nobis. ... A radiant Transpersonal ASELF walked into the Circle of the I AM, and henceforth recognized herself as the ONE WE ARE.

Notes

1. In MPD, the composite "I" becomes a composite "we," which incorporates as well as forms around a core of (more or less) complete, distinct personalities. Each *alter* has a characteristic cognitive style, religious and moral beliefs, specific food preferences, its own history, recollections, and circle of friends. Alters often dramatically differ from one another. Amazingly, some alters test positive for diabetes, while others *belonging to the same person* do not. Similarly, some, but not all alters, have allergies or need eye-glasses. Alters are teaching us about the incredible range of skills and abilities inherent in the human psyche. Because many alters instantly change their appearance as well as their state of health, they imply that such abilities might be consciously learned and utilized by greater numbers of people. Despite its severe pathology, MPD points to greater flexibility in the organization of our individual personalities, as well as insight into the nature of our Psyches, and into reality itself. (The new DSM IV diagnostic criteria, effective 1995, refers to MPD as "Dissociative Identity Disorder.")
2. *Mother*: traditional feminine path, but projecting through ASELF. *Father*: Passive, silent, and angry at life (or mother, another projection). It is beyond the scope of this book to develop these ideas further. The interested reader might enjoy Bowen's work on Family Systems and Harriet Goldhor Lerner's *The Dance of Anger*.
3. The discipline of Transpersonal psychology, known as the Fourth Force of psychology, incorporates the wisdom of many fields, including the humanities, the sciences, anthropology, medicine, mythology, spiritual traditions, and philosophy.
4. There are numerous levels to the unconscious. They include, but are not limited to, the personal, cultural (patripsych), global, collective, and transpersonal. Different levels of the unconscious direct and reflect Psyche's development during the cycle of life and include: the ground, archaic, submergent, embedded, and emergent levels of the unconscious. For a more in-depth description, please refer to Rowan's work on "Subpersonalities," pp. 145-148.

Selected Bibliography

Aaron T. Beck, M.D., *Cognitive Therapy and the Emotional Disorders* (New York, New American Library, 1976).

Robert Bly, *Iron John: A Book About Men* (Addison-Wesley Publishing Co., Inc., Reading, 1990).

John Bowlby, *Separation: Anxiety and Anger* Vol. II (New York, Basic Books, Inc., 1978).

Sukie Colegrave, *Uniting Heaven and Earth* (Los Angeles, Jeremy P. Tarcher, Inc., 1979).

E. Geleer, A.E. McCabe & C. Smith-Resnick, *Milan Family Therapy* (Jason Aronson, Inc., New York, 1990).

Harriet Goldhor Lerner, *The Dance of ANGER* (Harper & Row Publishers, New York, 1985).

Harriet Goldhor Lerner, *The Dance of INTIMACY* (Harper & Row Publishers, New York, 1989).

Jean Houston, *Public Like A Frog: Entering the Lives of Three Great Americans* (Quest Books, Wheaton, 1993).

Harold I. Kaplan and Benjamin J. Sadock, *Comprehensive Textbook of Psychiatry*, 4th Ed. (Baltimore, Williams and Wilkins, 1983), Vols. I and II.

Lawrence Le Shan, *How to Meditate: A Guide to Self-Discovery* (Boston, Little, Brown & Company, 1974).

Margaret S. Mahler, Fred Pine & Anni Bergman, *The Psychological Birth of the Human Infant: Symbiosis and Individuation* (New York, Basic Books, Inc. 1975).

John Nelson, *Healing The Split* (Los Angeles, Jeremy P. Tarcher, Inc., 1990).

Erich Neumann, *The Child* (New York, Harper & Row, 1972) pp. 10-11, 24-25, 109.

Erich Neumann, *Origins and History of Consciousness* (Princeton, Princeton University Press, 1970) p. 112, 416.

Erich Neumann, *The Great Mother* (Princeton, Princeton University Press, 1972).

Richard Noll, *The Jung Cult: Origins of a Charismatic Movement* (Princeton University Press, Princeton, 1994).

Samuel Osherson, *Finding Our Fathers* (Fawcette/Ballentine, New York, 1986).

Daniel V. Papero, *Bowen Family System Theory* (Allyn and Bacon, Boston, 1990).

John Rowan, *SUBPERSONALITIES: The People Inside Us* (Routledge, London, 1990).

Rudolf Steiner, *Cosmic Memory* (New York, Rudolf Steiner Publications, 1971).

Rudolf Steiner, *Theosophy of the Rosicrucians*, Lecture XII (London, Rudolph Steiner Press, 1907), pp. 128-129.

Rudolf Steiner, St. John Lectures (Cassel edition), 1909, p. 187.

Brian Swimme & Thomas Berry, *The Universe Story* (New York, Harper-San Francisco, 1992).

Jenny Wade, *Changes of Mind: Evolution of Consciousness* (S.U.N.Y., New York, 1996).

Christopher Wills, *The Runaway Brain* (New York, Basic Books, 1993).

D. W. Winnicott, *Playing & Reality* (London, Routledge, 1986).

Philip G. Zimbardo, *Shyness* (Jove/Addison-Wesley, New York, 1985).

We are leaving the dimensions of mind, thought, and feeling behind, to explore the more rarefied Realms of Spirit. Our experience within these sacred domains is but a glimpse and an echo of that which cannot be fully contained in words or concepts. We can never fully comprehend the unknowable. Do not be surprised by the ambiguous, poetic expression of that which can only be known and experienced directly.

The Realms of Spirit

Section five

> *"Go—not knowing where. Bring – not knowing what.*
> *The path is long, the way is unknown."*[1]
> —*Russian Fairy Tale*

The Path of the Heart

This section calls to the awakening Spirit. It invites you to suspend the critic and experience feelings and ideas that flow beneath words. It encourages you to get wet, cold, lost, and found, and to embark upon a most sensual and sacred journey. The Path to SPIRIT is the Path of the Heart and the doorway to self-realization and enlightenment. The journey upon this path reclaims our deeply rooted humanity and reaffirms the sacredness of all life.

The Realms of Spirit are the hallowed ground which creates and sustains ALL THAT IS. Throughout millennia, these sacred levels have been known by many names. Earlier, I introduced them to you as the Evolution of the Psyche (Section III), the Spectrum of Consciousness, the Great Chain of Being, and the Holomovement. Remarkably, new paradigms of science and psychology also describe levels within levels, fields within fields, and worlds within worlds, thus affirming the wisdom of ancient mystics and seers (see Table 5). A core of enduring wisdom, the Perennial Philosophy,[2] is embraced by all spiritual and mystical traditions (see Table 4). These ideas are already an integral part of a perennial psychology and perhaps, one day, will be included in a perennial science. Join me on this most sacred journey to the hallowed ground of ALL THAT IS.

Table 5
Path of the Heart

Four Worlds of Mystics	Great Chain of Being	Psyche	Consciousness
Origination	SPIRIT	Universal	ALL THAT IS/*SPIRIT*
Creation	Theology-spirit	Transcendent	SOUL/SELF
Formation	Psychology-Mind	Transpersonal	PSYCHE
Manifestation	Biology-Life	Existential	Self
	Physics-Matter	Ego	

As we travel through infinite Realms of SPIRIT, we grow, we develop, and we evolve into a richer understanding of Self and cosmos. We consciously participate in the sacred act of creation and begin to perceive reality in strikingly different ways.

The Yin and Yang of Perception

To embrace the wisdom of the mystic, we must suspend our rational way of understanding, and instead adopt a more intuitive, receptive mode of perception. We are asked to suspend our logic and reasoning skills. Intellectualization and thinking actually impede our exploration of inner dimensions. To engage these levels, we must leave behind our present beliefs, attitudes, and ideas, and shift into a different perceptual mode and state of consciousness. Shifting from analytic to receptive modes offers us the possibility of beholding the wondrous ground beneath all existence. We seek to encounter SPIRIT directly and gradually to attune ourselves to see, hear, touch, and recognize the Divine in everything we do. Our direct encounters with the Numinous always impart a felt sense and experience of expanded awareness.

There are two principle modes of perception; each bestows a different view of reality. One is rational and analytic, the other intuitive and receptive. The first allows us to survive in the physical world; the second connects us to SOURCE and helps us explore the Path of the Heart. These seemingly opposite modes of seeing, understanding, and relating represent a fundamental cosmic duality, and symbolize different stages in the evolution of the psyche. Yin (feminine and passive) and Yang (masculine and active) represent the negative and positive duality within all existence. Cosmic order reflects the ever shifting balance between the two principles. Refer to Figure 4.

Figure 4 reviews the bi-modal organization of Consciousness. The instrumental mode acts on the rational physical world of the senses, while the receptive mode assesses non-material, inner dimensions. Each impart a different form of consciousness and experience of Selfhood. Both are necessary for wholeness. Figure 4 integrates Western psychology with teachings of ancient spiritual traditions and audaciously offers a formulation of consciousness.[3]

The first, the mode of rational, objective reality, is understood through measurement, analysis, reasoning, and logic. It is the composite view of the left brain and the physical senses, as well as the microscopic and sub-quantum worlds uncovered through human technology. Rational Self, living and acting in the world of space-time and matter, perceives itself as separate and distinct from the rest of existence.

Figure 4.
Bi-Modal Model of Consciousness

SPIRITUAL HYPOTHESES

Mystics from markedly different cultures and epochs make similar assertions about spiritual consciousness:

1) Ordinary consciousness is narrow in scope and illusory in content, leading to a false belief about the nature of the self.
2) It is possible to experience the transcendent reality that underlies appearances. This reality is characterized by unity, purpose, and positive values.
3) The higher knowledge cannot be communicated in the terms of the ordinary world, but it can be experienced.
4) Attaining this experience requires the renunciation of self-centered aims and is aided by the practice of contemplative meditation and service.

If we consider the possibility that these assertions are valid, referring to another way of perceiving the world and ourselves, the question arises as to how we might understand the function of the recommended spiritual practices and the consciousness that results from them.

A first step is to consider that we have two basic modes of consciousness, depending on our dominant intention:

INSTRUMENTAL CONSCIOUSNESS
(EXAMPLE: DRIVING IN HEAVY TRAFFIC)

Intent: To ACT On the environment
Self:
Objectlike, localized, separate from others
Self-centered awareness
World:
Objects
Linear causality
Consciousness:
Focal attention
Sharp perceptual and cognitive boundaries
Logical thought, reasoning
Formal dominates sensual
Past/Future
Communication:
Language
Neurophysiology:
Sympathetic nervous system
Left hemisphere dominates
EEG: Increased beta waves
Decreased alpha and theta waves

RECEPTIVE CONSCIOUSNESS
(EXAMPLE: SOAKING IN A HOT TUB)

Intent: To RECEIVE the environment
Self:
Undifferentiated, nonlocalized, not distinct from environment
Blurring or merging of boundaries
World-centered awareness
Process
Simultaneity
Consciousness:
Diffuse attention
Blurred boundaries
Paralogic, intuition, fantasy
Sensual dominates formal
Now
Communication:
Music/art
Neurophysiology:
Parasympathetic system
Right hemisphere dominates
EEG: Decreased beta waves
Increased alpha and theta waves

FUNCTION OF SPIRITUAL PRACTICES

Overall Goals:
1) Subordination of instrumental consciousness to the receptive.
2) Deepening of receptive consciousness.

Contemplative Meditation

Meditation → Increased receptive intent / Decreased instrumental thought

1) Emphasizes attaining (receptive mode intent) rather than acting on (instrumental mode intent).
2) Focus on New (Receptive Time) versus Past/Future (Instrumental Time)

Renunciation of Self-Centered Aims

Renunciation → More opportunity for Receptive Mode perception

Decrease Survival (Instrumental) Self dominance.

Service

Service → Instrumental skills in the service of other-centered goals

Action for the sake of the task provides an alternative to material aims by providing a goal that focuses on the needs of others rather than oneself, opening receptivity to the environment.

SELF OF INSTRUMENTAL CONSCIOUSNESS

Characteristics:
Aim of self-preservation
Self as object
Separation
Effects:
Dissatisfaction
Traditional vices
Meaninglessness
Fear of Death
Importance:
Needed for individual survival

SELF OF SPIRITUAL CONSCIOUSNESS

Characteristics:
Permeated by environment
Resonant with environment
Self identified with larger life process
Effects:
Satisfaction
Expanded perception
Lessened fear of death
Meaning
Importance:
Needed for species survival

$$C = f(I + S)$$

CONSCIOUSNESS
INTENTION
SELF

Everything appears causally related and pre-determined. Time progressively flows from past to future, and we follow only one sequence of events at any given moment. Rational, intellectual mode predominates whenever Psyche is organized and ruled by the archetypal Patriarch. Our current world of separation and fragmentation is organized as well as governed by the Masculine aspect of the human psyche. Although the genius of the Patriarch has generated awesome technologies and inventions, it has also brought humankind to the very edge of chaos and self-destruction. Patriarch's insatiable need for power, control, and domination has led to war, crime, divorce, and desecration of earth. In order to achieve a more balanced perspective, humankind must open to another mode of perception and knowing.

The second perceptual mode is receptive and intuitive. It honors relationship, cooperation, and harmony, and desires to experience the unity of existence. Contemplation, introspection, and renunciation help guide consciousness to mysterious realms hidden from ordinary view. Receptive Self openly receives all of life's blessings. It allows a blurring and merging of boundaries to occur, and perceives everything as interconnected and interrelated within the tapestry of life. There is no separation; there are only wholes and relationships. Receptive mode identifies with larger life processes, simultaneous time, world-centered awareness, and the Sacred Circle, the circular-spiral matrix of SELF. Feelings, not facts, influence one's choices and actions. Intuitive Self seeks integration of mind, body, and spirit, and to become whole.

Receptive, Intuitive Self lives in the moment, trusting and accepting whatever is. This Self has no need to manipulate or control others. Instead, it seeks guidance and wisdom from within. A still mind and an open heart is all that is necessary for Intuitive Self to hear inspirational whisperings of SPIRIT. SPIRIT speaks to us in myths, stories, dreams, and nature. It coaxes us through music, art, poetry, waterfalls, starry nights, and ocean beaches. All are doorways to the Numinous.

Receptive, intuitive perception arises from the deepest core of our being; it is pure and untainted by earthly perceptions. Receptive mode, guided by the Feminine Principle, offers harmony and balance to rational, analytic Masculine. As masculine and feminine aspects unite, a richer and life-sustaining reality unfolds. Their Sacred Marriage heralds the emergence of Transpersonal specieshood and evolution of the psyche.

Spirit Speaks

SPIRIT is the language of Soul, spoken through the Heart.

To hear SPIRIT's Soulful call, we must first still our minds and open our hearts. SPIRIT's lyrical and poetic words entice us to awaken. We are summoned to inner worlds and dimensions beyond our earthly existence. Mystics and other explorers of these inner spaces use rhyme, metaphor, allegory, and verse to describe their direct encounters with the Divine. Their words burst with tantalizing hints of infinite levels within levels, and the unity of All That Exists.

Mystical passages frequently seem vague, ambiguous, and veiled to the uninitiated and unaware; often they appear to lead no where. But the words of the prophet reveal pearls of wisdom to those who listen with an open heart. Suspending your intellect will enable SPIRIT to reach into the deeper recesses of your mind. Know that these words are endowed with a higher wisdom, and let their significance slowly unfold and awaken the One We Are. Let us look at a few examples.

"The bamboo shadows are sweeping the stairs,
But no dust is stirred:
The moonlight penetrates deep in the bottom of the pool,
But no trace is left in the water."[4]

This poem relates to the notion of the "meritless deed" wherein an individual does something because is it appropriate, rather than for recognition, praise, or profit.

"The original reason of my coming to this country
Was to transmit the Law in order to save the confused;
One flower with five petals is unfolded,
And the bearing of fruit will by itself come."[5]

This Dharma (teaching) fortells the future development of Zen Buddhism in China. The "five petals" symbolize the five Zen fathers who recognized Zen as a distinct branch of Buddhism.

Spiritual teachers frequently answer a question with another. Important insights are usually concealed within each and every response. This process of uncovering one's own wisdom prepares the seeker for still deeper inquiry. On the other side of knowing, mystical encounters have an inner clarity and certainty that remain forever.

Stories and metaphors will be sprinkled throughout this section, since this is the way SPIRIT likes to teach. However you hear these words, I ask you to trust that deeper portions of yourself are already resonating with the wisdom that travels beneath these words. Stories, myths, and legends connect and unite humankind with the sacred ground of existence, and enable us to reclaim our spiritual heritage and roots. We return to earthly life more whole, more sane, and more fulfilled.

The three stories that follow speak of invisible worlds, hidden meanings, and the need to travel the Path of the Heart. If you take your time and allow the full meaning and significance of each tale into your heart, you will see that the three stories contain all you really need to know about the Realms of SPIRIT. The rest of this section explores these ideas and realms more deeply.

Story 1 – A Lesson in Receptivity

When I was a high school student, I was required to take a class in geometry. I did not like the subject matter at all; I found it very difficult and alien to my ordinary way of thinking. I struggled, but still was unable to understand all the many theorems and proofs. Often, I was filled with despair and frustration.

My teacher encouraged me to be patient. He told me to trust "the process" and said that the "curtain" would lift one day. He promised insight and mastery would eventually be mine. Month after month went by, and still the curtain did not lift. I continued to wait, to struggle, and to worry. The end of the semester was approaching and I had just about given up. Then, one morning as I awakened, I noticed a shift in my perception. Geometry was suddenly making sense. I was relieved; I had learned to perceive the world in a radically different way.

The quest for spiritual realization and enlightenment bears a striking resemblance to the process I experienced while studying geometry. The harder I tried to understand, the worse the results. Conversely, whenever I stopped resisting, things got better. I discovered that geometry was not a course I could learn with more practice or rational effort. Something new was required, and no one was able to tell me how or what I must do. The most anyone could do was encourage and guide my process and support my efforts. I chose to trust that learning was actually taking place beneath ordinary, rational consciousness, even though I did not know how. It would take time for the process to unfold. I was learning that transformation and growth would not be hurried.

Spiritual growth, like geometry, is an inward process. It, too, requires suspension of our usual, analytic ways of learning. Adoption of an inner process that occurs beneath our everyday consciousness is essential if we are to engage the Numinous. Similarly, no one can really explain to another how we find God; all we can ever do is share our own personal experiences and maps, and encourage others to continue their journey. No one can tell us how it is that we grow to 5' 2", or learn to walk, or learn geometry, or how we accomplish any of the miraculous and wonderful things we do each day. There are no simple formulas for solving our problems. Nor are there any guarantees that our journeys will be successful. Each of us must find our own particular ways to grow ourselves into the physical world. We must faithfully trust an inner wisdom, and allow SPIRIT into the deepest recesses of our minds and our hearts, if we wish to evolve spiritually. We must let go of analysis, we must stop trying and instead, invite SPIRIT to find us.

On our spiritual quest, we will know moments when enlightenment eludes our every effort. We will feel that nothing is happening and fear that the curtain of unknowing will never lift. We will suffer helplessness, confusion, and frustration. The processes of self-realization and evolution will carry us through many different feelings, experiences, and terrains. We will be constantly challenged to trust our journey and to continue our quest. And so it happens; without warning or design, and when the moment is exactly right, we hear, we see, and we touch the Divine. We know with absolute certainty; our way becomes more sure and we are transformed.

Story Two – The Path of Enlightenment

Every spiritual tradition and each person stands at the base of a very wide and very tall mountain. Each tradition and each Self travels along a unique pathway to the top of this mountain, seeking spiritual enlightenment and self-realization. Some roads are long, others short; some are easy, others quite arduous. The shape and extent of one's travels is known only upon our journey's end. The higher we ascend, the more expansive and revealing the view. Self's interaction with this sacred mountain is always transformative.

No two paths are alike. Some persons (and traditions) proclaim their path superior to all others; they will coax and urge you to travel along with them. Do not stray from your personal pathway once you have found it. Do not lose "your" way. At first, you may seem to travel alone. The higher you go, the narrower the mountain becomes. As you evolve and explore higher levels of Self, you notice the many separate pathways intersecting, overlapping, and finally merging. You begin to greet fellow travelers on the Path of the Heart. As you continue your ascent, all roads intertwine and meet until there is but one road. Finally, you reach the very top of this mountain and discover all of us are already there.

Story Three – The Veil of Forgetting

We are about to hear a story about the creation of an earth child, ASELF, who will live in New York City during the twentieth century. Sara and Victor will be the proud parents of ASELF and her two sisters. During this lifetime, ASELF will assist SOUL's exploration of three important themes: balance, evolution of the psyche, and a radical shift in women's awareness of themselves and their abilities. This lifetime will be filled with extraordinary challenges and possibilities.

Soul informs ASELF that her consciousness will descend and pass through the Worlds of Emanation, Creation, and Formation. She will follow the path of birth and creativity until she reaches her dwelling place in the World of Manifestation. Soul hopes that ASELF will one day spiritually awaken and re-trace this sacred path. ASOUL fervently wishes to embrace and unite with ASELF. Immediately after her birth, an angel gently places a kiss upon ASELF's cheek. Once kissed, ASELF no longer remembers how she came to be. And so it happens; ASOUL endows this newly born earth child with a spark of divine Consciousness. Instantly, ASELF awakens and her life mysteriously begins.

Perennial Philosophy

Ideas and symbols from the disciplines of science, psychology, and spirituality are presently converging. We are witnessing the emergence of scientific and psychological versions of the Perennial Philosophy, the fundamental core and essence of all spiritual traditions. Please refer to Table 4, page 86.

Perennial Philosophy tells us that Consciousness exists à priori, implying that It exists necessarily. Without this sacred ground, there is no existence. This Aware-ized Intelligence inherently yearns to create universes within universes, and to experience Itself in multiple forms and contexts. Each form of consciousness is endowed with a spark of the Divine. Hence, everything is an aspect of the Godhead.* All of existence has an in-built propensity for expression and creativity. Thus, we are all part of a multi-faceted, multi-leveled, multi-dimensional hologram projected into matter from ever more rarefied dimensions.

Various strata continually enfold and unfold, pulsating in and out of existence with each "present" moment. Higher levels explicate, project, and organize lower levels into holons.† Regardless of grade or

---

*Godhead refers to the essential being of God, the Supreme Being or deity.
†A holon is a whole at one stage which becomes part of a larger whole at the next.

degree, all realms and dimensions imperceptibly shade and blend into one another, from the most subtle to the most dense. Each realm is a different gradation of Absolute Mind. (These levels have been described as the Four Worlds, the Great Chain of Being, the Spectrum of Consciousness, and the Evolution of the Psyche.) A benevolent Awareness interpenetrates all levels of reality, establishing a unitive field in which all things are interconnected.

Essentially, all existence is a variation of One Consciousness.

Our personal position within this spectrum determines our understanding of Self, Universe, and God. Our understanding of divinity cannot be separated from who we are. Each consciousness perceives God in its own unique way. Humankind is invited to evolve and ascend through these multiple planes and dimensions in order to gain entry to other, more encompassing dimensions of awareness. Psyche yearns to travel the upward path from Self to SELF, so that we may embrace SOUL and unite with SOURCE. This is the path of spiritual enlightenment and the evolution of the psyche.

SOURCE descends into matter to experience Itself in multitudinous forms and contexts. It is simultaneously Observer and Observed. Each Self continually engages in the perpetual unfolding of infinite worlds and universes. We are the gods emerging; we are beginning to awaken to our spiritual heritage.

"Look inside, thou art Buddha."[6]
—Buddha

Psyche transcends heaven and earth. In its descent and involution into matter, it traverses hidden, rarefied dimensions alongside, beneath, and above our everyday world. These invisible realms organize and guide our daily existence. SOURCE generates, permeates, and sustains all of creation. All consciousness dwells within this sacred, holy ground. Whether one envisions this unitary field[*] of enfolded realms within realms as a holomovement, holoarchy, spectrum, ladder, circle, or spiral is not important. Each symbolically maps a pathway to the Numinous. All of our belief systems collectively reveal a vast, unmanifest territory in which Psyche, Self, and world are embedded and immersed. Humanity is suddenly re-discovering a fundamental, sacred unity and Its role within the conscious universe. We become ever more aware,

---

[*]Quantum physics considers fields as real. These non-material, non-visible structures within the fabric of space itself influence and organize the processes of creation and transformation.

".... that the world is uncreated, as time itself is, without beginning or end."[7]
—*Mahaparana (India, 9th C.)*

The parallel models of science, psychology, and spiritual traditions state that the phenomenal world arises from choices designed and created in upper orders of existence, and are later activated by an observing consciousness within the world of space-time.

"We do not 'come into' this world; we come out of it as leaves from a tree. As the ocean 'waves,' the Universe 'peoples.' "[8]
—*Alan Watts*

Many Mansions

The continuous processes of descent and ascent evoke the creation and evolution of a divine Consciousness as It actualizes Itself within multiple dimensions. As we awaken to higher consciousness, we realize untapped potentials and our true nature. SOUL's descent and involution into matter is the path of birth and creativity. SOUL heralds the initiation and genesis of new forms and new ideas. Our evolution to higher levels of consciousness is our quest for enlightenment and transcendence. This is the process of reclaiming one's SOUL.

Similar myths, legends, and stories of endless, hidden dimensions of consciousness beyond our physical universe are found in all mystical and spiritual traditions. Mystics penetrate into multi-dimensional worlds of consciousness and tell us the greater context wherein humanity's story unfolds. All traditions recognize the in-dwelling presence of SPIRIT deep within us. Myths and legends tell us how we are called, pulled, beckoned, and urged to higher states of awareness, and how we may become conscious co-creators of ourselves and our worlds.

"There is thus an incessant multiplication of the inexhaustible One and the unification of the indefinitely Many. Such are the beginnings and endings of worlds and of individual beings: expanded from a point without position or dimensions and a now without date or duration."[9]
—*Coomarasamy*

The Four Worlds

My version of the Four Worlds of Emanation, Creation, Formation, and Manifestion is a synthesis of the collective beliefs and ideas of many spiritual traditions. These ideas are presented in story form, because they allow me to communicate directly with your SOUL. Table 6 illustrates the relationship of the Four Worlds to SPIRIT, SOUL, PSYCHE, and Self.

Table 6
Mystical Four Worlds

Mystical	Consciousness	Attribute	Element
Origination/ Emanation Non-Being Genesis	All That IS/SPIRIT	Will	FIRE
Creation Life Themes Ideas Birth Creativity	SOUL/SELF	Intellect	AIR
Formation forms, molds shapes, options potential, choices possibilities	PSYCHE	Emotion	WATER
Manifestation action—events experiences situations objects	Self	Action	EARTH

However, the Four Worlds cannot be considered a true hierarchy or ladder. The concept of spectrum is also inaccurate because levels are not separate and distinct. At best, words can only partially describe the Realms of SPIRIT. Rather, everything is mutually interconnected and interpenetrated by SPIRIT, and every level feeds back to influence all other levels. The Four Worlds are part of an undivided, coordinated WHOLE called SOURCE. This concept is beautifully expressed by Wilber: "...The infinite is present in its entirety at every point of space, all of the infinite is fully present HERE."[10]

World of Origination

Each of us emerges from a state of NON-BEING. This is the mystical Void, the realm of No Thing, Non Self, and Emptiness. There is no beginning or ending; NON-BEING is a timeless, eternal state occurring in each present moment. It exists always. From this mysterious, dimensionless domain, time, space, and matter emerge. NON-BEING is the dwelling place of an aware-ized, benevolent ENERGY which lovingly creates infinite worlds, universes of the Mind, and levels of reality for SPIRIT to experience and explore. Some refer to this dimension as the World of Origination and Emanation. Within this domain dwells what we understand as "God, Source, Creator, The Divine, SPIRIT, Shiva, Brahman, and ALL THAT IS."

SPIRIT endlessly dreams unlimited possibilities and potentials into actualization. In this way, SPIRIT expresses Its boundless energy and creativity through myriad beings, worlds, and universes. ALL THAT IS wisely understands the rich diversities within unity and has conceived a wondrous plan. It would allow each of its imaginings to become a separate entity, apart from its Creator, knowing that each spark still remains a portion of ALL THAT IS. It sees no need to limit or control Its offspring. Rather, SOURCE lovingly bestows each being with its own unique awareness, allowing every aspect of The Godhead to experience Itself at the center of its world and universe. Thus, each consciousness is innately endowed with free will and desire to create Anew. And every form of awareness enfolds a spark of the Divine to span all of eternity. Simultaneously, each of us is an individual expression of divinity, as well as an aspect of a still greater whole. We are the dual reflections of the One and the Many.

With unbounded trust and love, SOURCE endlessly seeds its creativity into infinite psychic realms and universes. With a tremendous explosion, ALL THAT IS lovingly creates ALL THAT IS. And so it happens; A SOUL awakens.

World of Creation

Each world is interpenetrated and organized by the ones directly above it. Whereas SPIRIT dwells in the World of Origination, SOUL resides in the World of Creation. SOUL forms the core of our personhood. We know SOUL as our "higher" SELF, our greater potential. SOUL resembles and reflects its SOURCE. SOUL, too, creates offspring, endowing all portions of Itself with free will and consciousness. These Selves are born into physicality and the world of space and time. And so it happens; SOUL gives birth to you and to me.

Psyche is the spiritual messenger which mediates between SOUL and Self, enabling one to communicate with the other. Psyche is the thread connecting heaven and earth, as well as the spiritual essence which embraces and unites all aspects of our being. And so it happens; APSYCHE unfolds.

Creation hums with imagination and creativity. The many events that fill our lives and our everyday worlds are developed within this creative level. ASOUL decides the particular historic time period in which to seed a Self, and determines the various themes and challenges to be explored in any lifetime. ASOUL provides us with unlimited variations, choices, and probable enactments of our themes, granting each personal Self opportunities to freely choose circumstances with the greatest potentials for growth and for fulfillment.*

Hidden within the depths of the SOUL are the most profound mysteries of existence. As we journey from Self to SELF, we uncover our personal answers and wisdom. Gradually, we become whole. Through inner inquiry, we recognize the unity and connectedness of Self and Universe. As we explore infinite non-material realms of personhood, we become partners in the most exciting endeavor imaginable... Creation.

World of Formation

SOUL designs and generates multitudinous psychic blueprints, forms, archetypes, and probable events in the realm of Creation. PSYCHE deposits and stores these forms in the World of Formation. All orders of time and space are enfolded and exist simultaneously within the World of Formation. SOUL has an in-built propensity and yearning to translate Its ideas and desires into an infinity of forms and patterns; ASOUL directs, produces, and orchestrates our daily earthly experience, and allows ASELF to freely choose among an endless array of probable choices and opportunities.

SOUL whispers through our dreams, and urges us to heed its call. SOUL desires partnership with its multitudinous offspring, continually encouraging each to awaken to expanded, enriched Selfhood. SOUL urges us to travel the path from Self to SELF and life to LIFE. ASOUL and ASELF are each manifestations of the One We Are.

---

*These ideas are illustrated in later sections on Maya and Dreaming.

World of Manifestation

The World of Manifestation unfolds as SPIRIT continues Its descent into density and collapses into observable matter. This is the phenomenal world and the realm of earthly existence. It is a dimension in which the beliefs and dreams of consciousness are translated into practice. Within this world, consciousness freely chooses from infinite possibilities as it creates the events, circumstances, and experiences of our daily lives. And so it happens; ASELF is manifest.

The World of Manifestation and matter is one of particles and objects, events and situations. It is the world of the child, adolescent, family, home, and marketplace. It is also a world of hotdogs, pizza, baseball, and apple pie. The World of Matter is the realm of doing and thinking, of learning and applying the skills of our trade. In the material world, ASELF contentedly dreams of ALL THAT IS, even as the Creator dreams of ASELF.

Upon this fourth level of consciousness, ASELF encounters the totality of thoughts, beliefs, and choices she has projected and impressed upon the world of physicality. ASELF needs only to look out there, into the phenomenal world of space and time, in order to know herself. The manifest world is an outward reflection of the inner psychic levels and Ground that collectively informs, enfolds, and sustains each and every aspect of ASELF's life and personhood.

Let me share two stories which illustrate the path of birth and creativity through the Four Worlds. The first tells us about a couple who wish to have a baby. They psychically inform SOUL, in the World of Creativity, of their desire to become parents. SOUL agrees and instantly creates the child's blueprint and form, as well as determines her particular life issues and goals. These blueprints remain in the World of Formation until the child is conceived in the physical world. Over the next nine months, SOUL's psychic blueprints will be miraculously transformed into ASELF. And so it happens; ACHILD is born into the World of Manifestation.

In the second story, "The Birth of an Idea," SOUL places an idea for a luxurious office complex in an architect's mind. Ideas always emanate from the World of Creation. SOUL also places several mental blueprints in the World of Formation. If the architect decides to act upon this idea, a mental blueprint of his choice will be transferred onto paper in the World of Manifestation. The architect will freely expand and creatively embellish these blueprints with ideas of his own. Months later, in the World of Manifestation and Action, the architect builds a magnificent building complex of stone, brick, and glass. SOUL is like our Fairy Godmother: It lovingly bestows wondrous gifts and treasures upon us. SOUL grants our fondest wishes, fulfills our noblest dreams, and assists us in expressing our creativity in the world of earthly existence.

Awakening Spirit

SPIRIT descends into the material world to grow and to evolve. During this process, Consciousness progressively interpenetrates and traverses increasing levels of density. An essential part of ASELF's challenge in this world is awakening to her true nature and purpose by acknowledging she is an aspect and reflection of SOURCE. To do so, ASELF must begin her inner journey of ascent and transcendence through the Four Realms.

ASELF's task is to awaken and embrace ASOUL, so that observer and actor again become One. APSYCHE evolves by re-tracing this sacred, metaphysical pathway through the Four Realms of awareness. APSYCHE's evolution is the process whereby consciousness ascends and recovers wholeness and spiritual integrity.

Essentially, the Four Worlds span and form a spectrum of awareness and understanding from most subtle to most dense. Levels enfold within levels, ensuring that All existence is a boundless unity within the sacred grounds of ALL THAT IS.

> *"There is nowhere in the universe where one can go that is outside of MIND."*[11]
>
> —Wilber

Each realm reflects a particular level of evolutionary development and cosmic unfolding, and imparts a different grade of consciousness and expression to evolving PSYCHE. The journey of ascent and return parallels various developmental as well as psychological stages of Psyche as it heals, integrates, and transcends myriad levels of Ego, Existential, Transpersonal, Collective, and Transcendental Self, to at last realize Its eternal validity and unity with ALL THAT IS. It is each hero's and heroine's journey of initiation and return.

Pathways

ASOUL knew SOURCE created multitudinous pathways for personal growth and awareness. ASOUL was particularly interested in exploring the evolution of the psyche. Desiring conscious experience, ASOUL seeded numerous offspring into varying belief systems and civilizations throughout humankind's sojourn upon earth. Each tradition would serve as a university of the Mind, and each offspring would explore a different path to the One We Are. The experience of each pathway varies from individual to individual. Refer to Tables 7A and 7B.

Table 7A
Pathways—Correspondence with Four Worlds

Mystical	Kabbalah	HUNA	CHAKRA	PSYCHE
Origination	Atziluth	BEING	3rd Eye/Crown	Universal
Creation	Briah	DREAM	Throat	Transcendental
Formation	Yetzirah	Psychology	Heart Solar Plexus	Transpersonal
				Existential
Manifestation	Asiyyah	Matter	Spinal/Sacral	Ego

Table 7B
Pathways—Correspondence with Four Worlds

Amer. Indian and Mayan	Gt. Chain of Being	Doctrine of 5 Koshas Hindu and Buddhist	
	SPIRIT	BRAHMAN	CLEAR LIGHT
Assume personal authority	spirit/Soul theology—	vijnanas—6, 7 and 8/	transcendent
Dreamer and the knowing path	Psychology/ Mind	anandamayokosha/	transpersonal
Stop the World Death of the Advisor	Biology-Life	vijnanamayokosha manomayokosha pranamayokosha	Existential
Erase Personal History	Physics-Matter	annamayokosha	Ego

Offspring A lived in Safed during the 10th century A.D., and traveled the path of Kabbalah.* Within this system, the light of the Creator, Ein-Sof, the unbounded radiance of the Infinite, circulates through the Tree of Life and its ten energy centers, the Sefirot. Offspring A was particularly fond of Jacob's Ladder†, a cosmic map depicting four separate trees nested one within the other (see Figures 5 and 6). Each tree represents one of the Four Worlds of Emanation (Atziluth), Creation (Briah), Formation (Yetzirah), and Manifestation (Asiyyah). It is said that everything is contained within God Himself including the Tree of Life and the Four Worlds: "There is nothing outside of him." Ein-Sof is the "root of all roots" and the "cause of all causes."[12]

For the Jewish mystic, restoration and redemption of the world through Kabbalah is the path of the SOUL. Aware that "The World is a universe for the Soul,"[13] Offspring A was inevitably drawn to the Kabbalistic notion of three grades of SOUL: Nefesh, Ruah, and Neshamah.‡ The notion that each Soul-level reflects a different stage of personal evolution and spiritual awakening as one ascends through the four trees is unique to Jewish mysticism. The three Soul-levels form a tripartite unity (similar to ASOUL, APSYCHE, and ASELF) in which SOUL mediates between humanity and Ein-Sof. We must ascend and pass through gates of SOUL, in order to become the One We Are.

Offspring B was a Catholic prelate, who lived in Rome during the 17th century. He recognized the Doctrine of the Trinity and the Mystery of Christ as central to Catholicism. Offspring B was deeply impressed, as well as often perplexed, by the mutual indwelling of three distinct manifestations of one Godhead, each dwelling within the other.†† He acknowledged the divinity and evolutionary procession of God the Father, through Jesus Christ the Son, and the Holy Spirit. Offspring B considered Truth and Love as the essence of spiritual evolution. Upon learning that both Jesus and the Holy Spirit were sent

---

*Kabbalah means "to receive." Kabbalah is spelled differently in various spiritual disciplines. "K" is used in Jewish mysticism, "C" in Christian traditions, and "Q" indicates the esoteric Quabbalah.

†"Jacob's Ladder" symbolizes the interpretation and interlocking of the extended Tree of Life with the Four Worlds. "In this scheme, each World or level, with its own sub-Tree, emerges out of the centre of the one above, so that there is an interpenetration of Emanation, Creation, Formation and Action." Halevi, p. 41.

‡The new Kabbalah list five degrees: nefesh, ruah, neshamah, as well as two higher soul qualities of hayyah and yehidah, each belonging to one of five worlds. The world of the Godhead or ALL THAT IS forms the Fifth World.

††The unity of divinity, with one aspect dwelling within others, is known as "circumincession."

Figure 5
   Jacob's Ladder
       The Hierarchy of
       the Four Worlds

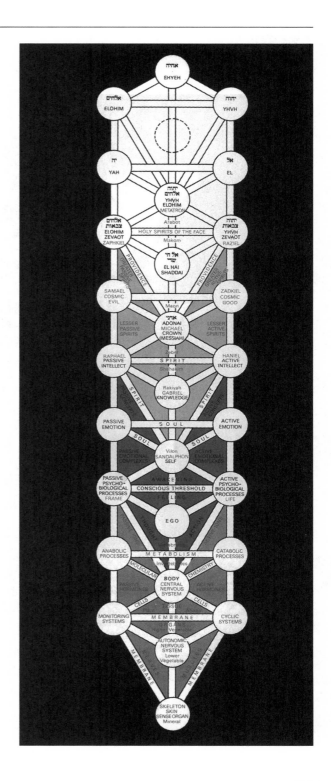

"The extended tree, which shows the interlocking of the Four Worlds.... Each World or level, with its own sub-tree, emerges out of the centre of the one above, so that there is an interpenetration of Emanation, Creation, Formation and Action. Thus, man, who has within him the four corresponding levels of divinity, spirit, psyche and body, can perceive all the Worlds in his inner and outer return to the Source...."

Source: Z'ev ben Shimon Halevi, *Kabbalah: Tradition of hidden knowledge* (New York, Thames and Hudson, Inc., 1992), p. 41.

Figure 6
Worlds within Worlds

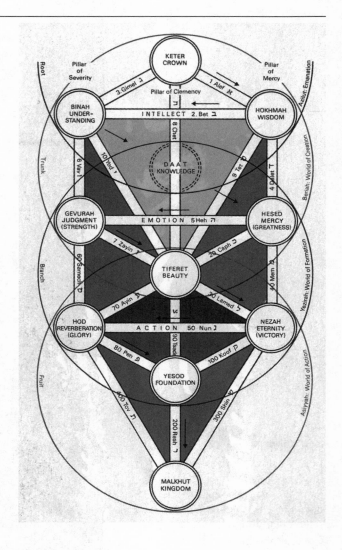

"The Four large circles show levels within a single Tree which correspond to the Four Worlds of Emanation, Creation, Formation and Action. The Sephirotic Tree is an image of Divine Unity."

Source: Z'ev ben Shimon Halevi, *Kabbalah: Tradition of hidden knowledge* (New York, Thames and Hudson, Inc., 1992), p. 40.

to earth to save and redeem humankind, this prelate openly wept tears of gratitude and joy. He believed the Word made Flesh, avowed God's divine presence in Christ and within our world. Jesus's new commandment became the guiding principle of his life:

"*Just as I have loved you, you also should love one another.*"[14]

In a moment of ecstatic revelation, Offspring B discerned the presence of God deep within his heart and in All of existence; He experienced Oneness with the Divine.

In another lifetime, Offspring C chose to become a Yogi and to study the Kundalini* path of the seven chakras. Kundalini Yoga unveils an ancient system of body-mind development and spiritual unfolding. The task of the Yogi is to stimulate an upward flow of transforming kundalini energy through the nadis.† As awakened kundalini energy flows sequentially upward from chakra to chakra, both physical and psychological unfolding ensue. Each chakra imparts its own particular perception and level of understanding and awareness. Offspring C consciously activated this chakra energy, as he visualized himself at the center of an eternal, dynamic mandala.

Offspring D became a Sufi mystic in the 13th century and explored the exquisitely beautiful inner gardens of the alam al-mithal. His journey allowed him simultaneously to dwell in seven inner worlds. With the help of his beloved, he reached the Garden of the Soul. Surrendering inwardly,‡ Offspring D realized the inner space of emptiness where Lover and Beloved unite.

Offspring E lived among the earth's First People and shared communion with the Great Mother. In her youth, Offspring E experienced a mysterious illness that became a point of entry and initiation into the Imaginal world of the shaman. Here, myriad potential events and latent possibilities awaiting actualization in the middle world of everyday existence are encountered. Offspring E became a "skillful technician of the sacred." Gradually, she learned to alter her state of consciousness at will. Through a shift in focus and intentionality, she gained access to information and knowledge not available in the middle world of space and time. Her journeys of ascendency carried her into the magical, expansive upper world where she received guidance and healing for others in her community. In contrast, her descent into the shadowy lower world was always fraught with obstacles and challenges that required resolution before lost portions of herself and others could be retrieved. Offspring E became a great visionary and seer who could discern the future shape and course of world events. Sometimes she was even able to intercede and alter destinies on behalf of her tribe.

---

*Kundalini is the personal and feminine form of a fundamental life force, *prana,* that pervades all of creation. Its other names include ch'i, ki, and the holy spirit.
†The nadis correspond to acupuncture meridians.
‡Surrending inwardly or Faña is a process of total annihilation and losing one's Self in God.

The following myth is found in various forms among the First People who still live in the Americas:

"This is a beautiful world, the Gods said, but we must never forget that it is made of the four previous worlds of spirit, mind, emotion and action. It is the man-animals work to hold the four previous suns together within himself in a balanced manner: the fire of spirit, the air of the mind, the water of the emotions, and the material of the earth."[15]

Offspring F lived among the Mayans in the Americas. He dwelled in the world of form and manifestation known as the Tonal, and dreamed of exploring the mysterious, unknown Nagual. He became a student of an old Indian sorcerer, who revealed the five ways one transcends the Tonal to reach the Nagual. (These five steps, uncovered millennia ago, correspond to the egoic, existential, transpersonal, and transcendental levels described by Transpersonal psychology.*) Gradually, Offspring F learned to erase personal history and transcend the personal "I" of the ego. Next, he endured the death of the advisor and discovered the interconnectivity within wholeness, as he approached the levels of Existential Self. Offspring F slowly learned to stop the world of the senses and to glimpse the creative energies he embodied. He gradually reached transpersonal realms of awareness. Many years later, the dreamer and the knowing path began to unfold, permitting Offspring F to see the manifest world through pure awareness. Returning from moments of ecstacy and union with ASOUL, Offspring F assumed authority and responsibility for all he experienced. He had encountered the awesome realm of the unknowable.

16th century Hawaii was the beautiful dwelling place of Offspring G. Huna cosmology also describes Four Worlds. The first realm, Being, is a holistic dimension, while the second World of Dreams is a symbolic one. The third, a Psychological domain, is subjective in nature, whereas the fourth is our ordinary World of Earthly Existence. In this lifetime, Offspring G chose not to become a Kahuna. Instead, he experienced only the world of earthly life, although he knew of the existence of other dimensions.

Offspring H was born in India during the 17th century. As he grew and matured, he passionately embraced ideas of Vedanta Hinduism. He grew aware that every individual is ensheathed in five koshas, each corresponding to a particular level of awareness. Offspring H understood

---

*The path through the Tonal, to gain access to the Nagual, parallels the ascent and evolution of Consciousness through the four worlds, as well as the psychological evolution of psyche through the "Spectrum of Consciousness."

The Realms of Spirit

that the five koshas spanned a continuum of gross, subtle, and causal states, and recognized that five koshas and three manifest states collectively engaged waking, dreaming, and non-dreaming levels of consciousness. Offspring H became immersed in a multi-faceted, multi-dimensional system of awareness. As he journeyed through these many levels, Offspring H began to peel away different koshas, one by one, allowing more subtle consciousness to emerge. He progressively ventured through the outer, lower sheath of annamayakosha (physical), through pranamayokosha (body), then manomayakosha (rational, abstract mind), as well as vijnanamayakosha (higher, subtle mind), until he arrived at the most inward and highest level of anandamayokosha (transcendental, spiritual bliss). With great assistance and encouragement from ASOUL and APSYCHE, Offspring H successfully navigated the Great Chain of Being and infinite realms of ALL THAT IS. He had realized the teachings of his tradition and touched his own SOUL. See Figure 7.

Figure 7
Comparison of Spectrum of Consciousness and Vedanta Psychology (Hindu)

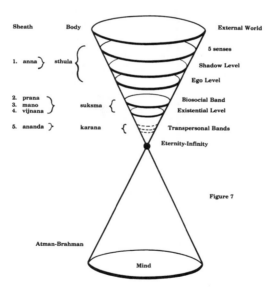

"This Figure shows the correspondence of the 8 vijnanas of Hindu Vedanta psychology with the psychological levels of The Spectrum of Consciousness."

Source: Ken Wilber, The Spectrum of Consciousness (Wheaton, Theosophical Publishing House, 1989) pp. 168–170.

Once Upon ASOUL

An eleventh century Tibetan, Offspring I chose to follow the eightfold path of Buddha's Dharma, whereby self-frustration and suffering (Dukkha) is brought to an end. He worked to heal the internal dualities and opposites within his psyche. Offspring I sought to maintain conscious awareness of his passage through six bardos states of waking, dreaming, meditation, dying, after-death, and re-birth. Bardos states span a continuum of consciousness connecting many, many lifetimes. Exploration of the eight vijnanas, the eight levels of consciousness, enabled Offspring I to enter the heart of SPIRIT. Though he struggled valiantly to realize NON-SELF and to reach CLEAR LIGHT, he died before fully uniting with ASOUL. See Figure 8.

Figure 8
Doctrine of Five Sheaths (Koshas)

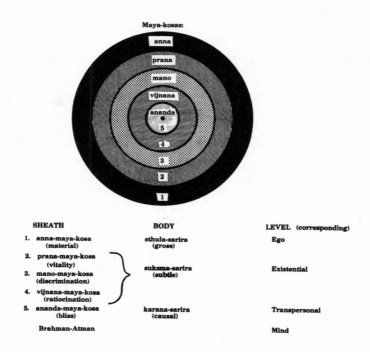

"The doctrine of Five Sheaths (Koshas) of Vendanta Psychology, Tibetan Buddhism and its Correspondence with the Spectrum of Consciousness."

Source: Ken Wilber, *The Spectrum of Consciousness* (Wheaton Theosophical Publishing House, 1989), pp. 176–177.

In summation, the mystical Perennial Philosophy forms the core of all major spiritual disciplines. Each tradition embues these ideas with its own unique perception of ALL THAT IS.

New Paradigm scientists have recently offered their own strikingly similar model of infinite invisible levels, strata, and worlds beneath Earth. The narrative below, based on the ideas of Bohm, Dyson, Everett, Herbert, Hiley, Peat, Sarfatti, Sheldrake, Wheeler, Wolf, and others on the cutting edge of physics, is my understanding of science's emerging paradigm of the Four Worlds. Like Bohm, I call it "Holomovement."

Holomovement

The Holomovement is a vast ocean of energy filled with Intelligence and Awareness. It creates and informs all. Holomovement resembles the mystical Void and the World of Origination. It also exists necessarily. The Holomovement is genesis; it forms an undivided, unbroken, seamless whole in which all existence dwells and is sustained. See Table 8.

Table 8
Four Worlds of Mystic and Scientist

Mystical	Science	Role
Origination	Holomovement	Non-Being, VOID, Ocean of Energy, Intelligence Awareness, SOURCE
Creation	Super-Implicate	creativity, ideas themes, information
Formation	Implicate	patterns, forms, shapes, choices, M-fields, options, possibilities orders of space-time
Manifestation	Explicate	action, experiences, situations, events, objects, matter

Before the beginning and before time, space, and matter, there was but a vast sea of Light, an Aware-ized, Intelligent energy. It remains unnameable, undefinable, ineffable, and unformed. For some unknown reason, a small ripple appeared upon this ground of all existence. Suddenly and tumultuously, this ripple exploded into a wave pulse that brought the physical universe and all that exists into being. An infinite spectrum of orders and levels of Intelligence and Awareness become manifest. Perchance other ripples and other Big Bangs also occur. Perchance other universes and realities also emerge from this immense ocean of LIGHT. Perchance an even more immense ocean of energy exists, and perhaps. ... And so it happens; the Holomovement appears.

SOURCE creates infinite levels and orders of existence, from most rarefied to most dense. All levels emerge from the Holomovement and all are enfolded in its unity. It is the Originator of Everything. Through the processes of involution and projection, information and meaning are enfolded into limitless levels of the Holomovement, to form Super Implicate and Implicate Orders. And so it happens; within deep levels of the Holomovement, a SUPER IMPLICATE level is organized.

Super Implicate Order

The Super Implicate Order is also an invisible realm of undivided wholeness and intelligence. Multiple worlds and realities are conceived and born here. Only the eternal "present" exists in this non-linear, timeless dimension. The Super Implicate levels are domains of great creativity and enormous potential. The Super Implicate supports, organizes, and develops Implicate levels below.

Super Implicate is a dynamic realm of formative cause. All causation takes place outside of space and time, through the continual processes of enfoldment and unfoldment. On this inner level, the evolution of meaning and the unfolding of consciousness contemplates form and pattern. Consciousness creates endless forms that await humankind's invitation into the manifest world. Super Implicate designs and generates morphogenetic fields, the quantum potential, archetypes, blueprints, life scripts, guiding principles, and patternings of our everyday world. An aware Intelligence designs and injects this information into the level below, where it will await its potentiation into matter. And so it happens; an IMPLICATE level is formed.

Implicate Order

An earthly observer must participate and interact with the passive Implicate Order before events and objects can manifest in the physical world. As part of an enduring cycle of birth and death, all physical matter appears and disappears, as it continually unfolds and enfolds within the vast depths of the Implicate.

The Implicate is also an order of unlimited levels, as well as a storehouse of infinite probable events. An observer engages this dimension in order to select the events and situations he or she will later encounter in physical life. This is a realm in which all orders of time are unfolded and explicated from still deeper, more subtle levels of the Implicate Order. In the Implicate, everything is interconnected through non-locality. All events, lifetimes, and worlds of space and time exist at once, in "potentia." When summoned forth by consciousness, time sequentially unfolds into the material world, moment by moment. Everything that ever has been, is, or ever will be, exists within the Implicate. All events passively await an invitation from an observer in the Explicate, before they may unfold into matter. Worlds pulsate in and out of existence and miraculously project into time and history. And so it happens; AN EXPLICATE order unfolds.

Explicate Order

This is the world of time, space, and materialized objects. Consciousness and Intelligence circulate throughout, informing every dimension of existence. An earthly observer witnesses the unfolding of the universe and allows our physical world to materialize. We are the actors. We choose and initiate a particular sequence of events and actions in the Implicate so that "we" may engage and experience it in the Explicated world of space-time. Hence, the actor is both creator and participant. Explication infers an audience. The phenomenal world arises from choices designed in the Super Implicate region, stored in the Implicate level, and chosen from the Explicate. And so it happens; AWORLD is manifest.

I like to use the electron as an example of implication and explication. When one electron in a pair changes spin, the Implicate Order instantaneously directs its twin electron to also reverse spin. Despite the enormous distances between them, both electrons in the pair reverse their spin at precisely the same instant. Non-locality requires that another level, beyond space-time and matter, orchestrates the events and happenings of our everyday world. Information from the Implicate instantaneously guides and directs each electron's trajectory and spin in the physical world. Mental and physical levels are reflections of one reality; both elements are necessary to the unfolding of our world.

Another story: Wishing to design a new dance score, a choreographer accesses the creative energies of the Super Implicate level. A rich interplay and exchange of information among many levels of existence ensues. Gradually, the dance takes form and shape. The score tells the dancers where to stand and how to move. As earthly dancers begin to interpret this score, they freely improvise and explore the dance's potential. This feedback helps the choreographer improve and refine her score still further. And so, a pre-existent score, modified and influenced by its creation (the dance), generates a cycle of creativity leading to still more dances. Essentially, the Holomovement is a psychic agency or field which creatively forms, informs, interacts with, and ensures the moment to moment unfolding of earthly existence.

Since there are a multiplicity of Four World paradigms, each with their own symbols and language, I will adopt the convention of referring to the Four Worlds only as Worlds of Origination, Creation, Formation, and Manifestation throughout the remainder of this book, for the sake of simplification.

A Round Trip

Each of us descends through infinite Realms of SPIRIT so that we may experience ourselves in physicality. But we must also be willing to ascend and re-trace this same pathway through the Four Worlds, if we wish to grow and evolve to our highest potential, individually and collectively. Our mission is to evolve and to return again and again, to these hidden, rarefied realms of existence so that we may heal, become more whole, and reclaim ASOUL. Our evolutionary journey ascends and descends a multiplicity of inner and outer dimensions of existence. In truth, the spiritual journey is a cyclic one. The Path of the Heart travels up and down, inward and outward, forward and backward, and all around. All directions are sacred pathways to the One We Are.

Descent into Matter

Consciousness freely descends into the World of Manifestation in order to realize Itself. As It descends into matter, it traverses the Four Worlds. This is the pathway all objects and events must travel in order to manifest. Consciousness gives birth to newborn infants, transforms architectural designs into buildings, creates symphonies, cascades over waterfalls, roars like a lion, and births great civilizations as It interpenetrates and creates all of existence.

Ascent and Evolution

The pathway of re-conciliation and union with SOURCE is also the journey of self-realization and enlightenment. It is the gateway to the Numinous. Humankind has encountered the blissful experience of Communion with the Great Mother during childhood, and the angry Separation of adolescence under the rule of the Patriarch. We have explored Ego and Existential consciousness, and are beginning to question earlier perceptions of Self and reality. APSYCHE seeks to heal and transcend itself. As consciousness achieves higher levels of organization, feminine and masculine elements will reconcile and unite in Sacred Marriage. A Transpersonal humanhood is emerging.

In summation: The pathways of ascent and descent reveal a continual process of transformation and enlightenment. I have previously referred to these pathways as the Great Chain of Being, the Evolution of the Psyche, the Spectrum of Consciousness, and the Four Worlds. If we wish to engage these in-dwelling psychic territories, we must first acknowledge their presence and existence. If we wish to consciously activate these realms, we must first understand their purpose and develop needed skills and tools. If we wish to gain access to other modes of knowing and comprehending the One We Are, we must shift our focus from ordinary to non-ordinary states of consciousness. As we undertake the journey of ascent and evolution, individually and en masse, we will experience more satisfying, fulfilling lives. And as we reach our sacred core, we will know our rightful place and time in the cosmos.

Let us briefly explore the concept of simultaneous time, which tells us that all of time and space exist in the Spacious Present. As we ascend to ever higher levels of awareness, we discover the timeless, eternal world of the mystic, the infinite invisible Realms of SPIRIT.

Simultaneous Time and Multiple Worlds

> *"There was a young lady girl named Bright,*
> *Whose speed was far faster than light,*
> *She traveled one day, In a relative way,*
> *And returned on the previous night."*[16]
>
> *—Buller*

Time emerges out of a realm that is itself, timeless. Each present moment is a bridge that unites a remembered past with an anticipated future. Our sensory and neurological apparatus continually translates all information and experience into a linear sequence of events that seem to take place "out there" in space and time. Collectively, time provides a contextual frame to our earthly lives and imparts meaning and significance to all we endure.

But each singular event, when viewed from other dimensions, is seen as an aspect of a still larger pattern. This pattern is a field of endless potential; it includes every conceivable variation of form this same event might assume under different circumstances. Hence, every particle and event exists first in potentia, within the World of Formation, until chosen for actualization in our world.

"The universe is pre-selected by consciousness."[17]
—*Wheeler*

ASOUL designs and injects multitudinous choices for ASELF to experience during her lifetime. Through her participation, ASELF evokes a potential that is already present in the World of Formation; her choice allows a Universe to unfold. Although she is not the sole creator of her reality, ASELF's cooperation is crucial to its realization. Universe and ASELF are partners in the sacred act of creation.

A cord of wood piled next to a fireplace provides a simple parallel. Matches have been placed on the mantle. Yet nothing happens until an observer decides to place the wood in the fireplace and ignite the logs. An observer transforms a latent possibility into a dynamic, glowing, cozy fire.

We are beginning to realize that connections between events and time do not take place "out there" in the world of physicality. Instead, everything first interconnects non-locally, in the subtle realms of the Implicate Order until selected and experienced by us.

Several scientists suggest that events not chosen for actualization in our earthly world are instead experienced in parallel worlds lying alongside our own. Cosmologists speculate that infinite universes and worlds come into being with every choice each of us make.* Recent theories of higher dimensional space (Hyperspace) allow scientists to explain and integrate seemingly divergent concepts into a "unified theory of everything."

"Listen, there's a hell of a universe next door: let's go!"[18]
—*e. e. cummings*

The concept of parallel and multiple universes raises some intriguing questions. Who or what is selecting the other variations of events that occur within these parallel worlds? Are these parallel versions of

---

*"The Dreaming Universe" by Fred Wolf provides further explanations and numerous examples of parallel universes.

you and me? Are all these variations of Joyce interconnected somewhere? How do I distinguish the "real" Joyce from the rest? How do we fit together? Do all of "us" belong to the same SOUL? Our individual answers are uncovered as we journey upon the Path of the Heart.

The notion of parallel universes, both physical and non-physical, shakes the very foundation of our identity as well as our understanding of the One We Are. As we contemplate these incredible ideas, our perceptions of reality and Self will profoundly alter, leading us to a radically different world-view and sense of humanness.

Mystics tell us that each moment of time occurs in the Now. All of existence and time are simultaneously present within the primordial ground of divine Consciousness; all pasts, presents, and futures are enfolded in the hidden realms of Source.

"There is but the one history and this is the soul's."[19]
—*Yeats*

ASOUL seeds each newly minted offspring into a discrete space-time sequence, hoping to provide each singular Self the most favorable opportunities for personal growth and evolution. From ASOUL's perspective in the World of Creativity, all offsprings and worlds exist in the eternal present, and all lifetimes occur at once. APSYCHE'S task is to mediate and communicate with all aspects of our collective "Family of Selves." In this way, APSYCHE deftly weaves all aspects of our SELFHOOD into the tapestry of life, ensuring that each individual's awareness and accomplishment blend and merge within the depths of ASOUL. In this way, the fulfillment of one Self is shared and becomes the fulfillment of all Selves. No Self is ever alone or ignored; all are cherished and embraced by ASOUL.

The world-views of science, psychology, and spirituality are blending. There is growing recognition that an Aware-ized Intelligence is the creator of Universe and ALL THAT IS. We suspect that each of us is a microcosm of a vast invisible macrocosm which lovingly organizes and sustains earthly existence.

We dwell in a participatory Universe which invites the involvement and interaction of ALL THAT IS. SOURCE creates so that It may know and experience Itself in myriads of ways. Universe awaits our choices and actions, individually and collectively. As humankind unites heaven and earth, AUNIVERSE unfolds. Without humanity's participation, time, space, and matter would not exist. A sacred Consciousness has endowed our species with incredible opportunity and awesome responsibility. Humanity and Universe are partners in the cosmic act of creating one another.

Many Lives

> *"Coming and going, life and death,*
> *A thousand hamlets, a million houses.*
> *Don't you get the point?*
> *Moon in the water, blossom in the sky."*[20]
> —Gizan

Spiritual traditions tell us that each individual lives many lifetimes upon our earth before attaining the enlightenment, wisdom, and compassion that release him or her from the reincarnational cycle of birth and death. Our lives follow one upon another, until we successfully master the lessons that enable us to complete this cycle. Humankind has descended into matter and the world of density in order to learn how to manipulate thoughts, ideas, beliefs, and desires in the material world of space and time. We must learn that the world outside faithfully reflects "who" we are inside. Everything we do, say, or think influences our earthly lives and incarnations. Our positive and negative choices are additive. Collectively they become our karmic debt to ALL THAT IS. We remain in the cycle of birth and death until we balance our Karma and learn the lessons of love, compassion, and forgiveness.

Karma and Reincarnation

Many traditions imply that Karma is a punishment; we reap whatever we sow. According to this interpretation, a future Self becomes accountable for the sins and failures of a previous Self, and a child is responsible for the sins of his or her ancestors.

In contrast, viewed from ASOUL's perspective, all our incarnations occur at once. Every lifetime becomes an opportunity for personal and collective growth and transformation. We are given as many lifetimes as we need to learn love and compassion. We cannot complete our work on earth until we learn to detach, let go, transcend ego, recognize our greater potential, and embrace ASOUL.

Manifest Self, living in the four-dimensional space-time world, "normally" perceives the passage of time as steadily flowing from one moment to another. But in truth, each discrete order of time that appears in the physical world is created anew, born out of the invisible ground of SPIRIT to which it returns. Viewed from higher levels of existence, all lifetimes unfold simultaneously.

Simultaneous time suggests a more empowering interpretation of reincarnation. Since all our lifetimes unfold at once, the notion of punishment is eliminated. Moreover, punishment is repressive; it diminishes creativity and action. Continual punishment leads to passivity and apathy; we simply give up and stop trying. Anger and rage begin to escalate, further prohibiting growth and spiritual transformation. Punishment results in a cycle of diminished ability and capacity, whereas the Law of Karma is intended to help us grow and evolve to higher levels of Selfhood.

Karmic cycles are opportunities to learn sacred truths, especially about Love. As we ascend the Great Chain of Being, we encounter only compassion for our struggles. As we reclaim feelings and attitudes mistakenly projected upon others, we personally experience love and wholeness. We are never blamed or judged. We are asked only to accept responsibility for our actions and the consequences that ensue from them. When we honor the Divine spark within all forms of consciousness, we are no longer able to violate another or to make war.

SPIRIT, through the Law of Karma, patiently provides us with as many lifetimes as we need to complete our journeys. We cannot hurry evolution; it unfolds according to its own rhythm. Collectively, all our simultaneous incarnations allow us to grow, develop, and evolve to deeper understanding and gnosis. This is the path of compassionate awakening.

Karma is an invitation to grow and evolve. The requirements for college graduation provide a useful analogy. In order to graduate, one must take a required number of classes in each of three categories: (1) a chosen area of specialization or "major," such as biology, psychology, chemistry, home economics, (2) areas of general education such as political science, history, English, mathematics, and (3) personally chosen classes (electives). Each course must be passed with a grade of "C" or better before one moves on to the next level course. Degrees are awarded only upon successful completion of all required courses.

The same is true with Karma. Once we enter the reincarnational cycle of birth and death, we must successfully master the many lessons of earthly existence before we may graduate. We cannot move to more complicated levels of experience before we acquire the necessary skills.

We remain within the reincarnational cycle until we understand that no one can gain at another's expense; until we align Self with the ethical and moral principles guiding Universe; until we accept responsibility for Self, our actions, and their consequences; and until we recognize Self to be part of a benevolent and sacred Ground that creates and sustains all forms of consciousness. The Karmic cycle of Reincarnation ends when each of us becomes the embodiment of love, compassion, and empathy.

Maya—World of Illusion

In essence, we are the sum total of all our many lifetimes. Every incarnation affects, as well as reflects, our continually changing perception of Self and reality. We have chosen to descend and to be born into physicality. Our incarnations are an opportunity to learn and to grow.

Yet our Spiritual traditions continually urge us to move beyond space-time and the material world. We are encouraged to transcend Ego consciousness and to awaken to our spiritual heritage. As we transcend Ego Self and ascend to other levels of awareness, we discover that the physical world is Maya, a world of illusion. We recognize and acknowledge the objectified, solid world we accept as "real" can only be perceived indirectly, through our sensory apparatus. We understand the illusory World of Maya is merely a stage upon which we act out our dreams and challenges, and live out our lives. Mystics and visionaries tell us that all of existence is *Energy*; Universe pulsates on and off with each moment's breath. These enlightened mystics remind us that everything participates in the continuous cycle of annihilation and creation. They also teach the impermanence of matter and advise that Consciousness, alone, is eternal.

Joe—A Game of Virtual Reality

It seems real-world experiences may be no more "real" than the virtual realities we generate on our computers. Virtual reality is a useful analogy that helps us understand how our multi-dimensional SELFHOOD unfolds into earthly existence. We watch as Joe slips a software disk into his computer and begins to interact with the simulated environment appearing on the screen. Within moments, Joe is so mesmerized, he has quickly forgotten this is a game. As he interacts with this mental "cyberspace," Joe reaps the consequences and rewards of his many choices. He wants only to enjoy the game and to learn whatever this challenging situation has to offer. He has no desire to make mistakes or to harm anyone. Sometimes Joe invites his friends to play along with him. Each player affects the choices and options of all the others and, as co-creators, all share in the outcome. See Figure 9.

Perhaps, as suggested by many spiritual traditions, most of us have mistaken an illusory world of Maya for the real one. Essentially, our SOULS have created and endowed us with a magnificent virtual reality and camouflage system. It seems so real that we, like Joe, have forgotten where we are. It is also possible that our universe, as well as each Self, is a multi-dimensional holographic virtual image explicated into the World of Manifestation from the Worlds of Creation and

The Realms of Spirit

Figure 9
JOE Playing a Game of Virtual Reality

Source: L. Casey Larijan, *The Virtual Reality Primer*, with Carl Machover, Ed., (New York, McGraw-Hill, Inc., 1994) p. 33.

Formation. We have been given the extraordinary privilege of participating with SPIRIT in the creation of our world and our lives.

Before Joe can purchase any equipment and enter the game, a computer engineer and software programmer must design and produce these products. This takes place in the World of Creation. ASOUL, assisted by APSYCHE, similarly designs and develops the necessary hardware, software, and blueprints for each Self's optimal growth.
In both instances, these components, patterns, and choices are deposited and await our selection in the World of Formation. Each of us has chosen to play an ingenious game called "Life." Our choices, decisions, and experiences are explicated from the World of Formation and unfold into physicality. Our position within the spectrum of awareness and our level of understanding determine the decisions we make.

All portions of your SELF are actively involved in creating the story of ALIFE. Certain aspects write the script, others produce and direct it. You, the actor, freely choose whether or not to play. You recognize that your choices and actions will affect all portions of your SELFHOOD. Your mission, then, is to acknowledge these other dimensions of your SELF, so that you may access their guidance and wisdom and engage the Numinous. In so doing, you unfold your greater potential and create a more meaningful life.

Each Self freely chooses whether to accept a particular challenge, and how to respond. We may blame, complain, deny, or avoid or instead, we may choose to engage fully with the experience in order to see where it will lead and what it can teach us about love and compassion. There is always deep meaning and purpose in our experiences.

Sometimes the events of our lives are exceedingly painful, devastating, and even horrendous. But knowing that each lifetime is carefully designed for our personal development enables us to co-operate with ASOUL and APSYCHE, rather than collapse into chronic despair and blame. Each of us must uncover the challenge and opportunity seeded within each situation, so that we may grow.

"Man's main task in life is to give birth to himself."[21]
—Fromm

Only with earthly experience can SOUL fully evolve. Actualizing our challenges in the physical world allows our higher SELF to put theory into practice. Only by experiencing pain, joys, stresses, and consequences that ensue from our actions and choices can we learn what our decisions are about. Physical life allows us to "walk our talk." As we act out the various events of our days, months, and years, we grow aware of the complexities and rich nuances within situations. If we knew with certainty that we would heal and become well, there would be no motivation or reason to change our ways and attitudes. It seems necessary to see, hear, feel, taste, and touch our experiences before we can fully recognize the One We Are.

The Art of Dreaming

"...dreaming is an extremely sophisticated art... to enhance and enlarge the scope of what can be perceived."[22]
—Castaneda

Whenever we dream, we leave our earthly bodies at home. We are freed from the constraints of ordinary space and time. Dream Universe is an amazing psychic realm that enables us to understand our pasts, to deal with our present, and to plan and rehearse our future events and experiences. It does so in extraordinary and innovative ways. In this magical world, we can explore ideas and potential decisions without fear of painful consequences. We can assess and evaluate potential situations before we make them "real." And we discover what is likely to happen if we choose to follow a particular event or decision into the manifest world. Dream Universe is an interface between inner and outer dimensions and a gateway to SOUL.

The Realms of Spirit

In the Dreaming realm, we are given the awesome privilege and responsibility of editing and rewriting our life script. We may even agree or disagree with ASOUL. This is the world of endless ideas, life scripts, blueprints, creativity, and all that we might potentially experience and know in the physical world of space, time, and matter.

Australian Aborigines are true artists of the Dream World. They journey regularly to an invisible dimension known as "Dream-Time," which resembles the World of Creation. From this most sacred realm, SPIRIT literally dreams life and Universe into existence, moment by moment. The Act of Creation is eternal.

When First People practiced the shamanic art of dreaming, they passed through seven gateways, each opening to a different aspect of Dream Universe. Each successive level endows the dreamer with heightened awareness and greater control of his or her dream activity. Dreaming then becomes a door to myriads of unexplored realms as "real" as our waking life. First People tell us we can learn to structure our waking world and our earthly lives quite differently by awakening within our dreams.

Similarly, the sorcerer's Art of Dreaming* goes beyond ordinary dreaming, wishing, imagining, envisioning, and day-dream activity. This kind of dreaming taps into an internal process which allows humankind to perceive hidden worlds and realms of existence that lie beyond ordinary space-time. These worlds are as unique, real, and engaging as our own. Dreaming gives us the ability to recognize and direct dream energies. We follow and ride these currents of energy into other worlds, much as the surfer learns to ride the crest of an ocean wave.

Buddhists tell us there are two types of dreaming. The first, ordinary dream-reality, relates to personal experiences and events, as well as memories of past lifetimes. The second, which sometime originates spontaneously, most often ensues from long years of spiritual practice and clarity of mind. Profound wisdom and lucid dreaming are encountered in this state.

For Yaqui sorcerer Carlos Castaneda, dreams serve as a point of entry to other realms and realities. For lucid dreamer Stephen LaBerge, dreams provide an opportunity to safely live out his wildest fantasies and to star in his own dream creation. For a Dream Yogi, dreams are rehearsals for the after-death bardos and a reservoir of spiritual wisdom. For physicist Fred Wolf, dreams help him become

---

*A sorcerer is an intermediary between seen and unseen worlds; a practitioner and teacher of the art of dreaming.

ever more Self-aware. And for the Hindu, the ultimate dreamer "Vishnu" floats on the cosmic ocean, dreaming the world into existence. And so it happens; AUNIVERSE unfolds.

Dream Time

Dreams and visions can be profoundly transformative.

> *"Cherish your visions and your dreams as they are the children of your soul; the blue prints of your ultimate achievements."*[23]
> —Napoleon Hill

It was the end of a busy day and ASELF had grown very tired. She needed to sleep. ASELF craved rest and healing. Body's wisdom knew sleep was essential for health and well-being. Gradually, ASELF turned off the lights, withdrew attention from her surroundings, and closed her eyes.

Moments later, she began her nightly journey to an inner dimension which re-arranged events of her waking world in spectacular ways. As ASELF gazed through the eyes of her ASOUL, she saw all of time unfold, as past, present, and future events blended and merged in new and innovative ways. She experienced herself traveling over vast distances to exotic lands in an instant. She visited her deceased grandmother and enjoyed a moment with a Neanderthal child. Like a magic carpet, her ideas carried ASELF along to anywhere and everywhen.

Nightly, ASELF and her Universe dreamed themselves into existence. ASELF understood that dreaming was a very personal, private act. She was never more alone than while dreaming. Yet paradoxically, she felt closest to SOURCE and ASOUL during these times. ASELF also valued the reflective quality of dreams; she thought they were like mirrors. Her dreams unpretentiously and honestly revealed who she was, rather than who she wished to be. They indicated needed change and how she might grow.

ASELF recognized that beyond sleeper, id, ego, and superego, she had a Dream Self. She wanted to know where Dream Self went while she was awake. She wondered if one aspect of personhood might be on while another signed off. Because time was simultaneous, ASELF even considered that both might actually exist at once. She wondered whether it was possible for Waking Self and Dream Self to ever directly encounter one another.

APSYCHE introduced ASELF to interdimensional levels and other worlds where unexpected possibilities and potential solutions to her earthly problems were presented for her consideration. ASELF imagined she was in a store filled with lots and lots of wonderful ideas. She clapped her hands and gestured toward a new scientific solution. In a twinkling, her selection was gift wrapped. She would take it with her when she returned to the waking world at night's end.

ASELF was then given a glimpse of a future event. She was told that this event would not manifest in her waking world for two more months. ASELF was very pleased with these coming attractions. She knew that she had chosen well. Her choices would ensure that the coming year would be a happy and fulfilling one for her and for her family.

Beam Me Down

Humankind is reclaiming a larger Selfhood and evolving a more compassionate reality for all who dwell upon our planet. All forms of consciousness, from the manifest to the unmanifest, join in this endeavor. With Universe's help, we dream our Selves into existence. Together, we are co-creating and participating in a sacred virtual reality. Hopefully, we will awaken to a fuller humanity and to a deeper understanding of life.

Each of us, like The Star Fleet Federation from "Star Trek," has descended into physicality to learn how to create and how to love. We have allowed ASOUL to "beam us down" upon the surface of the earth, so that we may grow and undertake the evolution of the psyche. When we complete this particular lifetime, we will ask ASOUL to "beam us up."

Revelation

A personal experience that originated from deeper levels of SELF afforded me lessons in love and compassion. As a result, I experienced a quantum leap in understanding.

Several years ago, I was having great difficulty accepting my daughter's lengthy illness. Alexandra had been ill for several months. None of our efforts to effect a cure were working. I was filled with guilt, pain, and despair. I could not let go. One morning, when I was not quite asleep or fully awake, a vivid image of the Virgin Mary appeared before me. Her eyes were filled with great pain and sadness. I felt her agony as she watched her son slowly and painfully die. In the background, I could see Jesus upon the cross. The scene took place at dusk on a bleak, cloudy day. She could not help him. A part of Mary died along with him. As I watched the Roman soldiers crucify her son I, too, could hardly bear it.

In an instant, the scene changed. Now I viewed a radiant, calm, and ecstatic Mary. Jesus had died and fulfilled his destiny. Mary had not intervened to save him. She understood his pivotal role in this historic passion play. This was what he was born to do. "It was done," she said. "He is at peace; he has completed his mission on earth." Mary was also at peace. Joy and serenity reached beyond her and into me.

I suddenly understood the message. It was encouraging me to let go and to honor Alexandra's journey and her experience. I had been politely told to get out of her way. This was perhaps the greatest lesson of love and compassion I have ever experienced. It enabled me to let go as well as to acknowledge my deep, abiding love for Alexandra. I moved beyond pain and sorrow to the realization of a mother's love. Months later, Alexandra began to heal and reaffirm her life.

As each of us begins to heal, we become more whole. As our sense of Self and personhood expand and evolve, we view events and situations from a deeper perspective and gnosis. Ideas we were unable to grasp weeks earlier, suddenly reveal themselves with amazing clarity, as we evolve and transcend Ego Self and our present world-views. We begin to understand events from a higher level of SELF. We glimpse a more compassionate reality and gain a fuller appreciation of our purpose. We return from our journey more loving and wise and henceforth, we know the One We Are.

Gnosis

Mystical experiences can reveal all of eternity in a flash. They fill us with awe and mystery and are usually transformative. But encounters with the Numinous are not always dramatic. Most often, they are quite subtle and gentle. Unless we pay attention to our everyday experiences, we will miss the many mystical opportunities which are offered us daily. The ordinary world overflows with evidence of the Divine.

By looking, we begin to see; by listening, we start to hear; and as we awaken, we discover the Spiritual Ground in which we dwell. By gradually attuning ourselves to SPIRIT's presence, we suddenly understand that SPIRIT is everywhere and everywhen, since it interpenetrates and pervades all of existence. This is the meaning of the well-known phrase, "Chop wood and carry water." These words tell us we can discover SPIRIT in everything we do and experience. SPIRIT reveals Itself in an infant's smile, a bird's song, a gust of wind, and a lover's touch.

William Blake shares his discovery of the Sacred in the Ordinary in this moving poem:

"To see a World in a grain of sand,
And a Heaven in a wild flower,
Hold infinity in the palm of your hand,
And eternity in an hour."[24]

Spiritual experiences, however they manifest, infuse us with gnosis. Through the act of "knowing," subject and object become one and all duality ceases. In a flash, we "know" absolutely. Encounters with the Numinous bestow an incredible certainty that our visions and experiences are very, very REAL. They remain with us Forever.

"The mystical is not how the world is, but that it is."[25]
—*Wittgenstein*

Realms of SPIRIT lie hidden beneath, above, alongside, as well as within the physical world we call "reality." These invisible, timeless realms design and create all that manifests into the world of physicality. SPIRIT interpenetrates and pervades all dimensions of BEING, ensuring that All of existence is interconnected and woven into the tapestry of life. Not only is the Absolute in all things, but all things are enfolded in the Absolute.

The upward journey from matter to non-matter and Self to SELF is an authentic path of spiritual awakening. The Path of the Heart is also the evolutionary journey of Psyche, in which Self strives to heal its wounds in its quest for wholeness. All paths ascend worlds upon worlds. Throughout this book, I have referred to them as the Spectrum of Consciousness, the Great Chain of Being, the Four Worlds, and the Evolution of the Psyche. Our understanding of Self, Universe, and SOURCE reflects our position within this spectrum of worlds within worlds and levels within levels.

Growing whole is as much a spiritual endeavor as a psychological necessity. The processes are essentially the same. As Self grows in awareness and gains clarity of purpose, it forges a deeper connection with SPIRIT and begins to unite with SOUL. We grow aware that humankind is deeply immersed in a game called "Life," and we understand that all aspects of our multi-dimensional SELFHOOD collaborate and create the events and encounters of our earthly existence.

In summary, all paradigms of the Four Worlds recognize an Aware-ized Intelligence as SOURCE and CREATOR of our Universe. Consciousness is the sacred ground of all existence. As SPIRIT descends and ascends in and out of matter, It travels the Path of the Heart. Consciousness lovingly creates worlds upon worlds for exploration and renewal. At the end of the journey of ascent and evolution, we experience the Sacred Marriage between Lover and the Beloved. Through Psyche's evolutionary journey, ASELF embraces ASOUL. At this moment, multi-dimensional SELF becomes whole. ASOUL, APSYCHE, and ASELF become ONE.

> *"We can only become what we are in the most natural core of our being, and just as a flower opens its petals to the sun, so does the seeker open his heart to the beloved."*[26]
>
> —Llewellyn Vaughn-Lee

Notes

1. Russian Fairy Tale, in David Schiller, *The Little ZEN Companion* (New York, Workman Publishing, 1994), p. 66.
2. Aldous Huxley, *The perennial philosophy* (New York, Harper & Row, 1945) p. vii.
3. Figure 3 was presented by Arthur Deikman, M. D., at the First American Interdisciplinary Scientific Conference on Consciousness, entitled *"Toward A Scientific Basis for Consciousness"* April 12-17, 1994 at The University of Arizona, Tuscon.
4. D. T. Suzuki, *Essays in Zen Buddhism* (First Series) (New York, Grove Weidenfeld, 1949), p. 352.
5. Ibid, p. 173.
6. Buddha: *Great Religions of the World* (Washington, D.C., National Geographic Society, 1971), p. 100.
7. Mahaparana (India, 9th C.) in Michio Kaku, *Hyperspace* (New York, Oxford University Press, 1994), p. 191.
8. Alan Watts in Charles Hillig (Ed.) *Cosmic Quotations* (Ojai, Black Dot Publications, 1993), Part III.
9. Ananda K. Coomarasamy in Ken Wilber, *Spectrum of Consciousness* (Wheaton, The Theosophical Publishing House, 1989), p. 14.
10. Ken Wilber, *Eye to Eye: The Quest for a New Paradigm* (Garden City, Anchor Press/Doubleday, 1983), p. 298.
11. Ken Wilber, *Spectrum of Consciousness* (Wheaton, The Theosophical Publishing House, 1989), p. 279.
12. Gershom Scholem, *Kabbalah* (New York, Quadrangle, 1974), pp. 90, 98.6.
13. Pearl Epstein, *Kabbalah: Way of the Jewish Mystic* (Boston, Shambhala, 1988).
14. Jesus, *New Testament* John: XIV, 3-5.
15. Peter Balin, *The Flight of Feathered Serpent* (Wilmot, Arcana Publishing Company, 1978).
16. A. Buller, Punch, in Michio Kaku, *Hyperspace* (New York, Oxford University Press, 1994), p. 233.
17. John Wheeler in Rick Fields with Peggy Taylor, Rex Weyler & Rick Ingrasci, *Chop Wood, Carry Water* (Los Angeles Jeremy P. Tarcher, Inc., 1984), p. 207.
18. E.E. Cummings in Michio Kaku, *Hyperspace* (New York, Oxford University Press, 1994), p. 217.
19. Yeats, "A Vision" in Jean Houston, *The Search for The Beloved* (Los Angeles, Jeremy P. Tarcher, Inc. 1989), p. 149.
20. Gizan in David Schiller, *The Little ZEN Companion* (New York, Workman Publishing, 1994), p. 290.

21. Erich Fromm in David Schiller, *Ibid*, p. 362
22. Carlos Castaneda, *The Art of Dreaming* (New York, Harper-Collins, 1993).
23. Napoleon Hill in *The Best of Success* (Lombard, Successories Publishing, 1994), p. BS 48.
24. William Blake in David Schiller, *The Little Zen Companion* (New York, Workman Publishing, 1994), p. 212.
25. Ludwig Wittgenstein, in David Schiller, *Ibid* p. 320.
26. Llewellyn Vaughan-Lee, *The Call & The Echo* (Putney, Threshold Books, 1992), p. 51.

Selected Bibliography

Jean Shinoda-Bolen, *The Tao of Psychology* (San Francisco, Harper-San Francisco, 1979), p. 89.
Barbara Ann Brennan, *Light Emerging: The Journey of Personal Healing* (New York, Bantam Books, 1993).
Fritjof Capra, *The Tao of Physics*, 3rd. Ed. (Berkeley, Shambhala, 1991), Chapter 12.
Paul Davies, *The Mind of God* (New York, Simon and Schuster, 1992).
Arthur J. Deikman, M.D., *The Observing Self* (Boston, Beacon Press, 1982).
David Deutsch & Michael Lockwood, "The Quantum Physics of Time Travel," *Scientific American*, March 1994, pp. 68-74.
Ken Dychtward, *Bodymind* (New York, Pantheon, 1977).
Freeman Dyson, *Theology and the Origins of Life*, lecture and discussion at the Center for Theology and the Natural Sciences, Berkeley, California, November 1982, p. 8.
David Fontana, *Secret Language of Dreams* (Chronicle Books, San Francisco, 1994).
Norman Friedman, *Bridging Science and Spirit* (St. Louis, Living Lakes Books, 1994).
Leonard George, *Alternative Realities: The Paranormal, The Mystic and The Transcendent in Human Experience* (Facts on File, Inc. New York, 1995).
J. Habermas, *Communications and the evolution of society*. T. McCarthy, Transl. (Boston, Beacon, 1979).
Z'ev ben Shimon Halevi, *Kabbalah: Tradition of hidden knowledge* (New York, Thames & Hudson, Inc., 1992).
Abraham Joshua Heshel, in Editors of the New Age Journal, *As Above, So Below* (Los Angeles, Jeremy P. Tarcher, Inc., 1992), p. 61.
Jean Houston, *The Search for the Beloved* (Los Angeles, Jeremy P. Tarcher, Inc., 1989).
David Allen Hulse, *The Key of it All* Vols. I and II (Llewellyn Publications, St. Paul, 1994).
Sir James Jean, in Rick Fields with Peggy Wood, Rex Weyler, and Rick Ingrasci, *Chop Wood, Carry Water* (Los Angeles Jeremy P. Tarcher, Inc. 1984), p. 207.
C. G. Jung, *Mandala Symbolism* R.F.C. Hull (Transl.) Vol. 9, Part I, Bollingen Series XX (Princeton University Press, Princeton, 1969).
Serge King, *Seeing Is Believing* pp. 44, 206, 222-223.
Serge Kahili King, *Mastering Your Hidden Self* (Wheaton, The Theosophical Publishing House, 1993).
L. Casey Larijani, *The Virtual Reality Primer* (New York, McGraw-Hill, Inc., 1994), p. 33.
Richard P. Mc Brien, *Catholicism* (San Francisco, Harper-San Francisco, 1981), pp. 343-365.
Caroline Myss, *Anatomy of the SPIRIT* (Harmony Books, New York, 1996).
Raimundo Panikkar, *The Trinity and the Religious Experience of Man* (New York, Orbis Books, 1973).

Charles Ponce, *Kabbalah* (San Francisco, Straight Arrow Books, 1973).
Karl Rahner (Ed.), *The Concise Sacramentum Mundi* (New York, The Seabury Press, 1975), p. 1764.
Peter Russell, *The White Hole in Time* (New York, Harper-San Francisco, 1992).
Z. Schacter-Shalomi with Ronald S. Miller, *From Age-ing into Sage-ing* (Warner Books, New York, 1995).
Adin Steinsaltz, *The Thirteen Petalled Rose* (New York, Basic Books, 1980).
Michael Talbot, *Beyond the Quantum* (New York, Bantam, 1986), p. 51.
Montague Ullman, *Dreams as Exceptional Human Experiences* in *ASPR Newsletter*, Vol XVIII. No. 4, pp. 1-6.
Jenny Wade, *Changes of Mind: Evolution of Consciousness* (SUNY, New York, 1996).
Roger Walsh and Frances Vaughan (Eds.), *Paths Beyond Ego* (Los Angeles, Jeremy P. Tarcher, Inc. 1993).
Michael Washburn, *The ego and the dynamic ground: A transpersonal theory of human development* (Albany, State University of New York Press, 1988).
Michael Washburn, *Two Patterns of Transcendence*, *Journal of Humanistic Psychology*, 30(3), pp. 84-112.
Alan W. Watts, *The Way of Zen* (New York, The New American Library, 1964).
Alan Watts, *Cloud-Hidden* (New York, Random House, Inc., 1974).
Margaret J. Wheatley, *Leadership and The New Science* (San Francisco Berret-Kochler Publishers, Inc. 1992), p. 59.
John White (Ed.), *Frontiers of Consciousness* (New York, Avon, 1974).
John White (Ed.), *Kundalini, Evolution and Enlightenment* (Garden City, Anchor Books, 1979).
Ken Wilber, *The Atman Project* (Wheaton, Quest Books, 1980).
Ken Wilber, *Up From Eden* (New York, Doubleday, 1981).
Ken Wilber, *Quantum Quotations: Mystical writings of the world's great physicists* (Boston, Shambhala, 1984).
Ken Wilber, *Sex, ecology, spirituality: The Spirit of evolution,* Vol. I, (Shambala, Boston, 1995).
Ken Wilber, *A Brief History of Everything* (Shambala, Boston, 1996).
Fred A. Wolf, *Taking the Quantum Leap* (New York, Harper & Row, 1989), pp. 80-81.
Fred A. Wolf, *The Dreaming Universe* (New York, Simon & Schuster, 1994).

The Realms of Spirit

We weave the many different dimensions and worlds together in the Circular-Spiral Matrix of Self. Each psychic principle guides a different evolutionary stage of personhood, individually and collectively. And every archetypal element endows a particular world-view which leads to a different understanding of personhood, family, relationship, economics, government, psychology, science, and spirituality. A humanity, dwelling again within a sacred circle of time, will experience Self and World as radically different. Let us explore some of these possibilities.

The Circular-Spiral Matrix

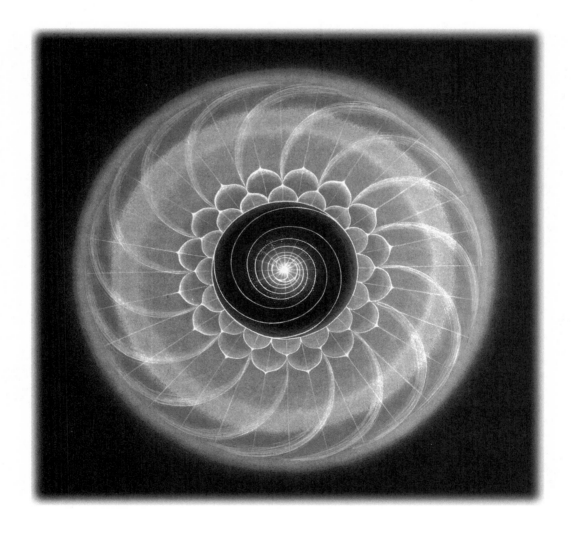

Section six

SELF was getting ready to celebrate her fifth wedding anniversary. It was hard to believe five years had passed since she married ENOUGH. ASELF smiled as she remembered her initial hesitation and fears. She had not felt ready to trust herself to another marriage. And yet she had been deeply in love with ENOUGH. ASELF was glad she had allowed herself to risk and to become vulnerable. Her second marriage had proven to be a happy, fulfilling one. This time she had chosen a worthy, loving husband for her lifelong partner and friend. ASELF and ENOUGH were still very much in love.

A year passed before the two families blended into one. ASELF recalled some of the tearful, angry confrontations they had experienced as each child uncovered his or her special place in this newly created family. ASELF knew that it was ENOUGH's patience and quiet strength that helped all of them through these moments. She watched as Harmony, Blessing, and Ananda grew into young adults, and realized that soon they would be off to college.

Father had moved to California following Mom's death. ASELF loved having him nearby. So did the children. He was a great help in bringing both families together. Father was everybody's Grandpa. Sadly, he had died two years ago. ASELF still struggled with her loss; she missed him dearly.

After Father died, ASELF moved more deeply into her spiritual practice. ENOUGH was also opening spiritually. They enjoyed sharing their inner experiences on the Path of the Heart with one another, and found that their marriage was stronger as a result. They both felt connected to SPIRIT and to ALL THAT IS, and acknowledged that each person and each form of consciousness has its rightful place in cosmos and on our planet. ASELF and ENOUGH knew and understood that All existence was precious and sacred.

ASELF recalled her internal fragmentation into a multiplicity of inner Selves during adolescence and young adulthood. It was then that ASELF had been figuring out who "She" was and how best to identify and follow her dreams. Why couldn't she be what she was "supposed" to be? Life could have been easier if she had been like her sisters. ASELF spent many years learning to recognize her projections and to take responsibility for her feelings and actions. With the help of Existential and Transpersonal Selves, she gradually integrated light and dark aspects of her personality, reclaimed her many disowned inner children, and released Ego's tenacious hold on Psyche. As she matured, ASELF grew more aware of her needs and gradually attained a sense of wholeness. Her marriage to ENOUGH had deepened ASELF's sense of inner value and contentment. Everyone watched her graceful transformation into a confident, caring, and most engaging woman.

Still striving for harmony in her life, ASELF was learning to balance her roles as wife, mother, dean, and Self. She wished for more hours in a day, but gradually she was becoming more accepting of her limitations. Her relationship with ENOUGH was founded upon the idea that neither would gain at the other's expense. It was important that each realize his or her talents and fulfill that potential as much as possible. And they would make time for family and relaxation.

ASELF's professional life was constantly changing. As Dean of Neuroscience, she spent little time in the laboratory. Even though she missed the excitement and mystery of research, ASELF considered her role as Dean an important one. It allowed her to shape and direct future research interests. She wanted to improve science's tarnished image and to raise academic standards. American research was lagging behind other countries. Foreign students previously had flocked to America; it was once considered a privilege to study here. Recently, ASELF noticed a sharp decline in numbers of foreign students. This decline paralleled reduced government funding for research and development, as well as lower academic standards and poor quality of education in our nation's public schools. ASELF thought it imperative for science to broaden its scope of investigation and to adopt a more inclusive research paradigm. As a Transpersonal scientist, ASELF was advocating for creation of a new scientific discipline, a Science of Consciousness.

Last year, Canada appointed ENOUGH their representative to the United Nations. He traveled continuously. ASELF gained a more global perspective of earth's problems as she accompanied her husband on many of his travels. She had a first-hand opportunity to see the extent and enormity of our planetary crises. ASELF grew more and more concerned each time she traveled. On one hand, science was responsible for wonderful discoveries and technologies. Everyone's life and daily routine was positively enhanced. On the other hand, she recognized the dark side of her world, her culture, and her life. She saw that the atmosphere was now dangerously polluted, holes in the ozone layer had enlarged, and toxic wastes continued to pour into rivers and lakes.

ASELF's world was in the midst of drastic change. Century-old institutions were disintegrating. Governments, as well as individuals, were declaring bankruptcy at alarming rates. Poverty was visible everywhere. Entire nations were starving and dying, while other nations were politically deteriorating and fragmenting. Ancient wounds and grievances were resurfacing and festering, while hatred and scapegoating were escalating. Riots and shootings were a daily occurrence; many nations were already engaged in war. ASELF felt the erosions of land and soil, as well as of values, ethics, morality, and human rights. Even the "family" was falling apart. She wondered what kind of world her children would inherit. She asked herself: Would the world survive until the 21st century?

Nations act very much like individuals. They, too, project problems and issues "out there" onto other nations, rather than confronting them "inside" their borders. Was it possible for countries to take responsibility for their attitudes and beliefs? While humanity possesses enormous skills and technology and is poised to explore the far reaches of our cosmos, we also face emotional, economic, and environmental crises of unprecedented proportion. ASELF was frightened; she recognized this to be the paradox of our present world.

Visions

ASELF awakened with a start. In just minutes, she had received an entire block of information about present global crises. She wasn't sure if she was dreaming or if she had seen a vision. Vivid, coherent, and persistent dreams had become a regular occurrence during her last three weeks. ASELF did not know who or what was sending her this information, nor what she was supposed to do with it. This dream ended with a biblical verse:

> *"And it shall come to pass afterward, that I will pour out my spirit upon all flesh, and your sons and daughters shall prophesy, your old men shall dream dreams, your young men shall see visions."*[1]

ASELF realized that her dreams were taking place within an invisible, hidden dimension, above and beyond time and space. At night, she would travel to this realm to develop gifts of prophecy and vision. ASELF would become one of many trained to assist humankind's transformational shift in consciousness. A Mystical Self was stirring deep within her psyche. Her visions and dreams afforded her a glimpse of evolutionary potential, and told her that a sacred opportunity for Spiritual birth beckons humankind. A new understanding of Self and Universe was emerging.

At first, ASELF's dreams reviewed the creation of Cosmos and earth, emergence of planetary life, and the remarkable gift of human consciousness. Subsequent dreams revealed many parallels between humankind's evolutionary stage of psychic development and the rise and fall of earth's cultures and civilizations. ASELF learned that Communion and First People symbolized humankind's childhood and life under the Goddess. She was then shown how humankind's adolescent quest for independence and control was a consequence of Ego/Patriarchal consciousness and denial of the Sacred. ASELF's final series of dreams revolved around existential anxiety and despair, present environmental crises, and the uncertainty of humankind's continued existence upon earth. Her dreams traced humanity's precarious survival to our lop-sided, distorted, and rigid views of reality and Self. She realized we had severed our connection to ALL THAT IS. ASELF discovered it was Ego's intense need to control and dominate the World of Manifestation that had brought humankind to the edge of chaos. Her dreams divulged the necessity for humankind to mature and to accept responsibility for its desecration of land, sea, and sky. Humanity must evolve or perish.

Just then the tone of ASELF's dreams and visions shifted. Now they were ever more hopeful. They encouraged humankind's acknowledgment and exploration of the Feminine aspect of Psyche. She was shown how a complementary partnership of feminine and masculine elements would restore balance and harmony to both inner and outer worlds. She was told that a Sacred Marriage between these two psychic elements would allow humankind to achieve its evolutionary potential and fulfill its promise. ASELF recognized Evolution as *THE* creative principle behind Universe, urging all consciousness to greater complexity and newly emergent, self-organizing properties. Her series of dreams concluded with an inspiring vision of a holy, sacred, and responsible humankind. Humankind was portrayed as a noble, peaceful People who dwell within the Sacred Circle, embrace SPIRIT, and recognize the unity of all existence. ASELF eagerly embraced her mystical lessons, since she recognized that her dreams were a special gift. She had recorded her dreams and reflections in her journal. Genesis is the first dream.

Dream 1—Genesis

PSYCHE approached a gathering of SOULS bearing a message from SOURCE. SOURCE was poised, ready to create a new Universe and to seed it with various forms of Consciousness and Intelligence. SOURCE was aware that many lifetimes would be necessary before consciousness would learn the lessons of this world.

SOURCE invited SOUL's participation in this exciting endeavor. Because SOUL is not able to manifest in physicality, it would be necessary to create multitudinous human offspring who could learn to manipulate thoughts, ideas, feelings, and desires in this dense plane of existence. SOUL's task would be the creation and guidance of humankind through space and time. Each SOUL voiced a passionate desire to be part of this amazing adventure. SOUL would nurture and prepare the sacred Consciousness that SPIRIT wished to bestow upon humankind. SPIRIT would support and guide the evolution of Universe. It would take billions of earth years to prepare an appropriate dwelling place for humanity. Only then could humankind emerge and embrace the ONE WE ARE.

SPIRIT intensely yearned to see this new Universe unfold. With great love, enormous creativity, and awesome inspiration, SPIRIT silently fashioned Universe out of no where, no when, and no matter. And with a spectacular flash of jubilant energy and celebration, Universe instantly burst into existence and formed everywhere and everywhen.

ASELF had just witnessed the moment of cosmogenesis. She watched as Universe exploded into existence 15 billion earth years earlier. This spectacular event endowed Universe with the miraculous gift of Life. The possibility of human existence was seeded in Universe's future. ASELF was deeply stirred by this cosmic dream.

Dream Two—Evolving Universe

Within this incredulously hot fire ball, ASELF observed a storm of particles, anti-particles, incipient matter, light, and energy moving in and out of existence for the briefest of moments. Nothing was permanent. Universe was becoming aware of infinite possibilities. Each primordial particle was given tremendous freedom to explore its potential and to uncover its unique identity. Particles of matter and anti-matter collided and danced with one another into oblivion, leaving a much smaller, finite number of particles to cooperate in the awesome task of creativity. Emergent Life arose from the larger whole called Universe. ASELF recognized that through this singular event, All existence was interrelated and interconnected.

ASELF wondered if Universe would collapse back upon itself, but this did not happen. Then she worried that Universe might expand too rapidly. She knew if this happened, matter would never coalesce and form. A very delicate and precise balance between gravitation and expansion occurred; circumstances favored particle interaction. Gradually, more order and restraints were imposed, as each momentary choice influenced as well as limited later ones.

Universe slowly cooled, allowing basic elements to form. Hydrogen was the first to appear. Helium, Carbon, Nickel, Iron, Copper, and many other elements followed. Galactic clouds formed and seeded small galaxies sprinkled with twinkling stars. Universe began to hum with creativity and to sparkle with Light. Small galaxies collided and formed still larger galaxies. ASELF watched with awe and amazement as a Star called Tiamet violently exploded into a spectacular supernova at the end of its life cycle; 4.6 billion earth years ago, Tiamet generously bestowed her cosmic gifts upon Universe. As ASELF continued to watch, she saw a solar disk develop from the remnants of this supernova. Our sun and its orbiting planets were created from the ashes of its predecessor, Tiamet. ASELF recognized the third planet from the Sun as humankind's future home. She suddenly realized that all planetary existence was the offspring of Mother Earth and Father Sky, and that we truly straddle heaven and earth.

ASELF had been made aware that a tremendous love and yearning pervaded Universe. She realized that eons and eons of time and preparation had been necessary before earth could become manifest. ASELF was beginning to see that every choice and interaction created a turning point and evoked a particular path of possibility and development for Universe to explore. Every*thing* was in relationship; No*thing* was separate. ASELF suddenly understood that Universe and its inhabitants always evolve together. Humanity and Universe Are One.

Dream 3—Life Begins

A week later, ASELF had another dream about awakening life. Earth initially experienced severe, millennia-long electrical storms as well as intense, searing heat. As earth gradually cooled down, its atmosphere, oceans, and continents formed. ASELF noted the triumphant emergence of earth's first ancestor, Prokaryote. Prokaryote[*] would successfully flourish for billions and billions of earth years.

---

[*]Prokaryotes are single-cell organisms without nuclei. They are earth's first life forms, and predominated for over 2 billion years. Some species still exist today.

Emergent life slowly mutated and differentiated into a multiplicity of uni-cellular forms. For almost two billion earth years, these primordial bacteria willingly cleansed our atmosphere as they dined upon a rich diet of complex organic compounds. So successful was Prokaryote's proliferation, so ravenous their consumption of earth's bounty, that alas, there were no longer enough nutrients and resources to sustain their existence. The process of photosynthesis also enabled Prokaryote to transform carbon dioxide, water, and Sun's golden light into oxygen. But the increase in oxygen levels proved dangerously toxic to Prokaryotes, and they began to die.

ASELF was wide awake; her heart was pounding heavily. She had glimpsed a frightening parallel between the Prokaryotes and present-day humanity. Would we, like our primordial ancestors, the Prokaryotes, proliferate beyond earth's capacity to feed and sustain us? Have we irreparably altered the conditions of our biosphere through continued pollution of air, land, and sea? Does humankind have enough time to reverse these conditions and ensure survival? She wasn't sure!

Dream 4—New Life Forms

After Prokaryote's demise, ASELF saw that ice silently covered much of our world; a long period of glaciation followed. Then a new cycle of Life began to unfold before her eyes. Nature brought forth a variety of plant and animal species that could thrive in our oxygen-rich atmosphere. The first of these new organisms, the Eukaryotes, appeared about two billion Earth years ago.* ASELF watched as specialized uni-cellular creatures combined and merged into various multi-cellular organisms. They learned to share their new abilities and talents with one another, and to cooperate in the recycling and balancing of air, land, and sea. Everything was precious; nothing was wasted. Nature's rhythms, seasons, and cycles offered a harmonious way of life to all of earth's creatures.

With great clarity and amazement, ASELF recognized that the evolutionary trend toward greater complexity, with newly emergent, self-organizing properties, was occurring on every level of creation. She had encountered this intrinsic principle during her personal exploration of the Spectrum of Consciousness, the Great Chain of Being, The Evolution of Psyche, and the Four Worlds. Now she was observing it in the World of Manifestation.

---

*Eukaryotes are uni-cellular organisms that contain a membrane bound nucleus, chromosomes, and DNA. Many forms reproduce sexually.

Dream 5—Evolution Accelerates

The pace of evolution and development was accelerating. Each complete orbit around our sun brought forth a greater diversity of life forms. Emerging life was successfully learning to live upon earth, as well as in her oceans. Flowers and trees now graced our planet with beauty as well as bounty. The relentless cycle of creation and annihilation throughout Universe continued. Mysteriously and suddenly, the dinosaurs disappeared. Now, mammals inhabited earth. The first of humankind's ancestors, Australophithecus Afarenus, was born five million earth years ago. Universe was filled with joy and excitement.

Again, the pace of evolution began to accelerate. An early, human-like species, Homo Habilis appeared nearly 2.6 million earth years ago and learned to fashion rudimentary tools of stone. By 1.5 million earth years ago, Homo Erectus, the hunter, roamed the African continent, and later migrated into parts of Asia and Europe. ASELF was told that Homo Erectus was the first life form to tame the forces of nature.

200,000 earth years ago, humanity's primordial ancestor, Homo Sapien, appeared. Other species, including the Neanderthal and Cro-Magnon, co-existed for a time, alongside early Homo Sapien. ASELF witnessed the gradual disappearance of these other species; only Homo Sapiens remained. Homo Sapiens possessed an extraordinary range of emotions and intelligence. Their language and communication skills were legendary. Earth's climate was again experiencing drastic change, forcing Homo Sapiens to migrate into North and South America, and as far south as Australia. Homo Sapiens would soon populate the entire world.

ASELF viewed the remarkable emergence of modern Homo Sapiens, 40,000 earth years ago. She knew that the hopes and dreams of humanity's future flowed through their veins. SPIRIT had waited 15 billion earth years to realize this dream. As ASELF watched, SPIRIT lovingly endowed each individual with a spark of Divinity. In a flash, Homo Sapiens were transformed into HUMANKIND. And so it happened; Humankind was born and Psyche began its evolutionary journey into space and time. Universe exploded with exhilaration and celebration. (See Table 9.)

Table 9
Evolution's Timeline

Cosmogenesis	15 Billion
Galaxies and Supernovas	10-14 Billion
The Solar System	5 Billion
Prokaryote	4 Billion
Eukaryote	2 Billion
Plants and Animals	350 Millions
Dinosaurs	235 Million
Birds	150 Million
Mammals	35 Million
Australophithecus afarensis	5 Million*
Humans—Homo Habilis	2.6 Million
Homo Erectus	1.5 Million
Clothing, Fire, Shelter	500 Thousand
Archaic Homo Sapiens, Cro-Magnon, Neanderthal	200 Thousand
Modern Homo Sapiens Humankind, First Peoples	40 Thousand
Classical Civilizations	3500 B.C.E.
Rise of Nations	1600 C.E.
Copernican Revolution Kepler, Galileo, Descartes	16th Century
Charles Darwin	19th Century
General and Special Relativity Quantum Physics Hyperspace	20th Century
Extinction or Transpersonal Specieshood	21st Century

Source: Briane Swimme & Thomas Berry, *The Universe Story* (New York, Harper-Collins, 1994) pp. 269–278.

---

*Scientists are beginning to discover evidence that hominoids may have dwelled upon earth far earlier than previously believed.

Dream 6—Physical Birth

ASELF's next dream revolved around themes of life and birth. She knew that all life begins within. A lengthy period of gestation would be necessary before consciousness, in its myriad forms, could manifest in physicality. Deep within every man and woman dwell precious seeds of humanity's future. At the moment of conception, egg and sperm enthusiastically embrace and unite; a tiny life begins to stir. For the next nine months, newly evolving life remains safely nestled within Mother's protective, nurturing womb. Evolving life is entirely dependent upon Mother for its every need. In reality, the two are One.

Physical Birth is a truly daring act of separation and trust; One now becomes Two. Although still totally dependent upon a Caretaker or Mothering One, the newly born must now also rely upon its own life support system for survival in an alien world. The daunting task of learning to thrive in the world of space-time and matter awaits. Mother's love and assistance were essential, as necessary as the breath of Life itself.

ASELF was suddenly overcome with sadness and concern for the growing number of earth's children who receive little parenting or guidance. She wondered if these children could develop the necessary skills to effectively participate in an increasingly complex and hostile world. And she worried about the effect millions of unparented, neglected, and abused children would have upon humanity's future.

Dream 7—First People

First People appeared on our planet 40,000 earth years ago, and blissfully dwelled within a Sacred Circle, an eternal, timeless NOW without beginning or end. ASELF realized that Mother Earth, Father Sky, and Humankind formed earth's First Family. She watched as Mother Earth embraced her newest offspring with oceanic love and adoration. All of humankind felt kinship with Universe, as it basked in the companionship of fellow plants and animals. Humankind belonged!

ASELF knew that First People were initially without tools, technology, and assistance, and that it would take many millennia to amass the knowledge and skills necessary for independence. Vulnerable, fledgling humankind made a Sacred Covenant to obey the rhythms, cycles, and rules of Great Mother. ASELF knew that Mother Earth and Father Sky would offer sanctuary and abundance in return, and a dwelling place wherein humanity could safely grow and evolve. Communion and peace prevailed.

ASELF intuitively understood that Communion is humanity's childhood. Throughout this period, the psychic archetype of Great

Mother/Goddess governs our perceptions of Self, world, and cosmos. When the Goddess reigns, Mother and child are One; Psyche and Universe form an undifferentiated, unbroken whole. Perceptively, Great Mother understood that Psyche, Earthling,[*] and First People would one day leave behind their pre-personal, passive, dependent consciousness of childhood and seek larger identities and Selfhood. To grow, all three would need to successfully navigate this critical developmental stage.

ASELF awakened from this dream, grateful to know First People would be nurtured and protected. She now suspected the tasks and challenges of Communion and Childhood were essentially the same. One described the evolution of a species, First People; the other portrayed the evolution of Psyche and Earthling through a Spectrum of Consciousness.

Dream 8—Psychological Birth

ASELF's next dream reviewed the *psychological* challenges of childhood and Communion. An in-built propensity and bias toward growth and creativity were urging Psyche, Earthling, and First People toward Ego consciousness. She was told that Psyche's ability to grow and evolve through the Spectrum of Consciousness was critical to the development of Earthling and First People.

ASELF dreamed that Psyche, Earthling, and Humankind had successfully resolved the Oedipal Crisis[†] and were soon to be initiated into the state of Ego/Patriarchal consciousness. Humanity would now experience autonomy and independent personhood. ASELF realized that individuation and conscious choice cannot occur without separation from the Matriarch. With great courage and determination, Psyche proceeded to *slay the dragon* and usurp the powers of Great Goddess.

ASELF learned that archetypal Patriarch would now govern Psyche, Earthling, and First People. Humanity had struggled and emerged triumphant. The remarkable task of *Psychological* birth had been accomplished, and Psyche, Earthling, and Humankind were now endowed with free will, choice, and responsibility. ASELF watched as humanity forged a unique peoplehood. Ego consciousness and the Patriarch were now their noble guides. As thirst for new horizons intensified, they proudly proclaimed their dominion over All.

---

[*]I will refer to each individual member of Humankind as "Earthling," in order to avoid confusion with ASELF, our heroine and dreamer. Note, ASELF is also an Earthling.
[†]Resolution of the Oedipal Crisis indicates a shift from the pre-personal, passive Matriarchal consciousness to Ego consciousness, separation, and individuation under the reign of the Patriarch.

Dream 9—Exodus

In her next dream, ASELF watched as humanity unfurled a separate, independent peoplehood and immersed itself more deeply into spacetime. As the Patriarch emerged, Psyche, Earthling, and Humankind renounced the Great Mother, repressed the emerging Feminine, disavowed the Sacred, and gradually severed conscious awareness of the inner realms of existence. Triumphant and invincible, Psyche, Earthling, and Humankind left the Sacred Circle and followed the linear arrow of time into a "Brave New World."

ASELF's dream indicated that an entirely new perspective and understanding of Self and Universe were unfolding. She witnessed Psyche's internal struggle for independence from Great Mother flowing outward into the World of Manifestation. New religions disavowed the Goddess and embraced the Great Patriarch and God the Father. Priestly Kings, all over the globe, claimed the powers of the Divine and rose to positions of dominance and control. ASELF realized that by emphasizing the dark, shadowy aspects of womanhood, "Good" Mother was transformed into "Bad" Mother, and Woman became the property of Man. ASELF suddenly felt a personal sense of loss, as well as great sorrow for all womankind. She understood that Ego consciousness had successfully usurped Great Mother's authority in both inner and outer worlds. Under the Patriarch, humankind forged a hierarchy of power and control, and arrogantly placed itself on top. Psyche, Earthling, and Humankind now embraced the Reality Principle and glorified the rational, linear, and logical mode of being.

Dream 10—Separation

ASELF's next dream revealed that humanity had chosen to develop its powers of rationality and reason while ignoring the emotional side of our nature. She learned that humankind would now search "outside" of itself for information and validation, accepting only objective, physical data that we can see, hear, taste, touch, and smell. Henceforth, we would deny and ignore the subjective inner experiences of our deeply rooted humanity. In this way, we would disown, split off, and repress the essential qualities of our unique specieshood. By the 20th century, Separation was complete. Humanity no longer seemed to miss the sacred. Science had become a powerful God.

ASELF felt great sorrow and despair when she recognized that humanity no longer knew or wished for Communion with nature. Tragically and defiantly, we chose to disregard the rules, rhythms, and property of Mother Earth. And so it happened; a Sacred Covenant was shamefully disavowed.

Dream 11—Warning Signs

ASELF's dream shifted its focus by disclosing a world on the verge of insanity and collapse. She knew that chaos and anarchy lurked in the shadows. ASELF was heart-broken as she observed humanity's desecration of earth's land, air, and sea. She wept as she watched thousands of starving refugees leave their country in search of sanctuary and food. She wondered how a "rational, intelligent" species could irresponsibly destroy its planetary home and kill fellow human beings.

ASELF realized that our everyday world was in the throes of uncertainty and chaos. Meaning, value, and purpose had sadly disappeared from the face of the earth; despair, apathy, and a pervasive sense of hopelessness saturated her world. ASELF recognized that human life and survival were in jeopardy. With profound sadness, ASELF realized that Psyche, Earthling, and Humankind had lost their way.

Dream 12—Crisis

ASELF's next dream indicated that a radical shift in consciousness was transpiring; "Good" Father/Patriarch was now transformed into "Bad" Father/Patriarch. No longer revered, science was becoming the scapegoat for all the ills and problems of humankind. Psyche, Earthling, and Humankind had become aware of Patriarch's dark side and unrestrained rationality. ASELF suspected that humankind had repressed, then banished disowned portions of its multi-faceted, multi-leveled Selfhood to Psyche's shadow.

ASELF was advised that to grow and evolve, Psyche, Earthling, and Humankind must correct their distorted, lop-sided views of reality and reclaim projections of superiority, arrogance, and domination that were responsible for desecration of Earth and Sky. Each must release duality, multiplicity, and psychic fragmentation, in order to attain wholeness and awareness. ASELF watched as the three moved away from a *black and white, either/or* perspective and instead, adopted a *both/and* mode of perception. Ego-centric Patriarchal consciousness was beginning to release its tenacious hold upon Psyche, Earthling, and Humankind. Ex-centric Patriarch would embrace a greater Selfhood and serve a more noble cause. Humanity was now standing before the gateway to existential awareness and in the midst of a third psycho-social stage of development. ASELF knew that if successful, humankind would experience profound transformation and realize a vastly different specieshood. She prayed for humankind to prevail.

Dream 13—Unfolding

A moment of profound decision was at hand: Humanity was confronted with its mortality and the need to choose its future course. Every human being would be asked to select and determine which 21st century, among an awesome range of probable futures, he or she wished to experience. ASELF hoped humankind would choose the Path of the Heart. She discovered that the necessary solutions to heal a world and to heal ASPECIES patiently awaited humanity's call.

Her dream indicated that humankind's total immersion into the depths of physicality and its repression of the sacred, were nearing an end. She was advised that the arrow of time somewhere and somewhen ceased to be linear; Psyche, Earthling, and Humankind now traveled upon a circular-spiral path. ASELF proudly watched the three claim their destiny at the interface between matter and non-matter. She realized a new era was unfolding. Her hopes for humankind are echoed in this ancient Ojibuay Prayer:

> *"Grandfather, look at our brokenness. We know that in all creation only the human family has strayed from the Sacred Way. We know that we are the ones who are divided and we are the ones who must come back together to walk in the Sacred Way. Grandfather, Sacred One, teach us love, compassion and honor that we may heal the earth and heal each other."*[2]

Dream 14—Circular-Spiral Matrix

"The circle is Great Spirit's emblem. All life is a circle."[3]

Her next dream revealed that SPIRIT is both form and ground; creator and created. Indeed, Consciousness contains and is All. ASELF envisioned Consciousness as a circular-spiral matrix of existence which yearns to experience ever higher levels of complexity, purpose, and meaning. She considered every life as an eternal breathing in and breathing out of a Consciousness which knows no beginning or end. ASELF noted a parallel between circular-spiral pathways of growth and transformation which begin and end in Psyche, and the circle without beginning or end. Higher and lower, better and worse, as well as right and wrong were no longer relevant upon transcendental levels of Awareness. Rather, unity and diversity, light and dark, matter and non-matter harmoniously co-exist within the Sacred Circle. See Figure 10.

Figure 10
Mandala—The Center of Self

Unity and diversity, light and dark, matter and non-matter harmoniously co-exist within the Sacred Circle.

Source: Jung - Picture 24

ASELF began to dream about the circular-spiral matrix. She was advised that images of circles, mandalas,* medicine wheels, and spirals often appear in dreams and visions. They symbolically proclaim Psyche's need for structure, harmony, balance, and new personhood. In truth, mandala and medicine wheel are sacred maps to the Realms of Spirit. The *Center* is nameless and eternal, a place of Self-renewal and rebirth. The *Center* is every Self's goal.

---

*The Sanskrit word *Mandala* means "circle." It delineates a sacred, protected space. It is a map of the cosmos. Jung considered the mandala as the symbol of SELF; simultaneously center, goal, and totality of Psyche. The American Indian Medicine Wheel symbolizes the holy Zero of Mathematics. Wheels are also sacred maps and mirrors of Creation, enduring life, emergent Self, and the realms of the Invisible.

ASELF recognized Evolution as a cosmic principle directing Universe toward greater creativity, complexity, and purpose. She was told there *are no absolutes*. Even Laws of Nature and Universe are continually transformed, as new levels of complexity and diversity emerge and unfold into the world of physicality. She had discovered a truly magical world of certainty/uncertainty, predictability/unpredictability, enormous possibility and paradox; she had discerned Universe's grace and majesty.

ASELF suspected that humanity's journey had always taken place within the Sacred Circle. Only humankind believed themselves to be outside and beyond the Sacred. ASELF realized the foolishness of Psyche, Earthling, and Humankind; she instinctively understood humanity to be exactly where it needed to be. Her dream disclosed that every human being enfolds a personal version of the Spectrum of Consciousness, the Great Chain of Being, and Worlds within Worlds. Psyche, Earthling, and Humankind have only to look inside to discover the One We Are. Perchance, SPIRIT's faith in humanity's ability to turn away from war, violence, greed, and destruction would ultimately prevail.

"We shall not cease from exploration,
and the end of all our exploring will be
to arrive where we started and know
the place for the very first time."
—*T.S. Eliot*[4]

Dream 15—Emergent Feminine

ASELF's dreams were revealing the larger context wherein humanity's drama unfolds. She was learning that Consciousness seeks fulfillment through a diversity of creativity and form, and discovered that all levels of Consciousness continually split off and *seemingly* disown portions of Itself. She knew that it had to do with "Letting Go," and allowing offspring opportunities to freely choose their own destiny and future. She understood that the gifts of free will and choice come with a mandate to accept personal responsibility for our choices. We cannot have one without the other; both are necessary for balance and wholeness. Psychological birth can only be attained through integrity and honesty.

The Feminine aspect of Psyche, Earthling, and Humankind was beginning to stir and awaken. ASELF remembered that the Feminine had been devalued and banished to Patriarch's shadow at the time of separation from Great Mother. Repression of the Feminine had offered the greatest opportunity for Patriarch's *initial* survival in physicality. But her banishment also brought separation, alienation, isolation, and existential despair.

ASELF's dream was urging humankind to reconcile the Feminine and Masculine principles in Psyche. The Feminine principle restores wholeness, harmony, and balance; she is healer, redeemer, and resurrector of SOULS. The Feminine was portrayed as trusting, intuitive, and receptive; her perception both direct and experiential. ASELF experienced an enormous sense of relief whenever she thought about the Feminine.

She then asked about the Goddess. ASELF was aware that many Earthlings wished to return to a simpler stage of dependency and passive protection afforded by Great Mother. Many were still running from responsibility and looking for some "other" to rescue and protect them. But ASELF's dreams insisted that humankind must not return to an earlier stage of development, nor confuse the Goddess with the emerging Feminine. To do so would invite serious psychological regression and disturbance. Goddess and Feminine were distinct psychic archetypes who preside over different developmental stages. Great Mother/Goddess governs childhood and youth, whereas Feminine brings forth mature, responsible adulthood. ASELF recognized the Feminine to be an interface between matter and non-matter. She serves as a gateway to the Numinous.

*Dream 16—Sacred Chrysalis**

In another dream, ASELF learned that Emergent Feminine heralds the awakening of Spirit, as Psyche, Earthling, and Humankind begin their sacred journey of ascent and evolution upon the Path of the Heart. ASELF realized that transcendence of the existential level of awareness and differentiation of the Feminine bestows Transpersonal Consciousness and a chance to embrace ASOUL. She was exuberant.

ASELF saw that all of humankind were enfolded within a sacred chrysalis which resides deep within the spiritual Soul of every human being. The sacred chrysalis affords humanity the safety and protection necessary to complete its evolutionary shift in consciousness. Sheltered within this intrapsychic space, the Masculine heals and matures; the Feminine grows and differentiates. ASELF was told that both psychic elements would dedicate their next stage of evolutionary development to the healing of Psyche, Earthling, and Humankind. Together, masculine and feminine principles would participate in the cleansing

---

*A chrysalis is a cocoon of loving security and safety. It provides an intrapsychic vessel for metamorphosis of Psyche, Earthling, and Humankind into Transpersonal Specieshood.

and replenishing of earth. ASELF's dreams were challenging each individual to evolve into a more compassionate, empathic, and benevolent specieshood. The beautiful poem below captures the experience of sacred chrysalis.

"CATERPILLAR DREAMS"

Floating in a hazy mist of finely woven gossamer,
With milky clouds and tatted lace so mercifully protecting my awakening.
Ripening in a hanging womb-like altar, swinging gently in the winds,
That are shaping and arranging what the inner worlds project into the ethers.
Here I sway, in stillness, in suspension, without mind, and yet I dream,
So snugly wrapped in mysteries, of moments in the splendor of awakening.
Until life becomes a prayer and inner knowing takes the place of outer wrappings.
The silken suit dissolves and melts to golden threads.
I shed them to the light.
Cocoons are but a memory, as faint as was the NOW that blossoms forth,
So skillfully recorded in the painted patterns of my splend'rous wings.
A living, breathing offering to others who lie wait in silken suits.
We all are an expression of the different stages of this thing called Life.
We are the breath of One who Loves, the heartbeat of a Universe Divine,
A breath well-breathed, desire relieved, an inhale or an exhale into time.

—Andrea Cagan[5]

Dream 17—Spiritual Birth

ASELF was pleased to learn that Psyche, Earthling, and Humankind were again a part of ALL THAT IS. All three acknowledged their sacred heritage and re-affirmed their hallowed covenant with Mother Earth and Father Sky. All vowed never to violate oath or Universe again. She watched a dancing Universe joyfully rush out and embrace humanity.

ASELF discerned that something else must happen if humankind is to realize its fullest potential. This was the meaning of the sacred chrysalis. She was shown the cycle of birth, death, and evolution of Psyche. She realized that each birth is a letting go and dying to a previous Selfhood. It is also an evolution to another level of awareness. Each level imparts new understanding and endows new personhood.

ASELF recognized that *Physical birth* is the culmination of long months of preparation and gestation. During this time, newly emergent Life gains the ability to dwell in the world of space-time and matter. Physical birth proclaims that Mind, Body, and Spirit have become One. Earthly life is its gift.

ASELF understood that *Psychological birth* is achieved only if the psychological challenges of early life are successfully resolved. It takes great courage to assert one's individuality and to create an independent Self. Psychological birth endows humankind with the gifts of free-will, choice, and responsibility, and grants the privilege of individual personhood to Psyche.

ASELF recalled that Emergent Feminine is a harbinger of *Spiritual birth*. Her awakening and evolution, in concert with growth and transformation of the Masculine, calls forth this sacred event. Spiritual birth takes place upon the Path of the Heart and generously bestows the gift of Transpersonal specieshood upon humankind. Transpersonal Human *is* co-creator of Self, world, and cosmos, and is joined in sacred partnership with ALL THAT IS.

ASELF learned that a segment of humankind would choose to evolve and ascend the Path of the Heart and the Great Chain of Being. She considered these individuals to be the seeds and future offspring of a new *Spiritual* humanity. Transpersonal human would soon dwell upon earth. She felt enormous gratitude as well as great sadness. Some, but not all, humankind would survive and flourish upon Earth. A portion of humanity would successfully complete the necessary tasks required to unfold and manifest a radically new world. As her dream ended, ASELF envisioned the birth of a *Spiritual* humanity. ASELF had witnessed the approaching shift from Existential to Transpersonal consciousness. And so it happened; TRANSPERSONAL HUMAN would dwell upon earth.

Dream 18—Sacred Marriage

ASELF watched Psyche and Humankind descend through the Four Worlds to dwell in physicality. She recalled the enchanted moment when ALL THAT IS endowed Homo Sapien with human Consciousness. She knew that it had been earth's most spectacular event. Yet, ASELF perceptively sensed something far more wondrous and mysterious about to unfold upon earth.

Once again, humanity was a part of a whole called Universe. Masculine had successfully embraced Ex-centric consciousness and let go of control and domination. He willingly supported the emerging Feminine. Feminine developed her precious gifts of receptivity, cooperation, and intuition, and proclaimed her psychic partnership with Masculine. ASELF knew that unity and reconciliation would prevail. Transpersonal Psyche, Earthling, and Humankind yearned for integrity and wholeness of the "Adult" kind.

ASELF discerned that Masculine and Feminine are not true opposites, although humanity had once believed them so. She was told that a larger view of reality is granted whenever these complementary archetypes unite. Enhanced perception, deeper understanding, and greater awareness are the gifts the two together bring. ASELF realized that the union of masculine and feminine principles allows the split in Psyche, Earthling, and Humankind to heal; Humanity becomes "I and Thou" as well as "We and Us." Wholeness and integrity are proclaimed and uncertainty, chaos, and turmoil left behind, as Masculine and Feminine embrace. She recognized that humankind now encompassed the whole of existence and dwelled in a magnificent technicolor world.

ASELF experienced a profound moment of understanding; Reconciliation and Union of masculine and feminine elements are the joining of intellect and intuition, object and subject, inner and outer, matter and non-matter. As Psyche, Earthling, and Humankind become whole, and Masculine and Feminine unite, the seeds of a new humanity are sprinkled throughout Universe. All are expressions of a Sacred Union. *ASACRED MARRIAGE* is a sacrament bestowed upon a loving, cooperative humankind as they heal themselves and their planetary home.

Psyche, Earthling, and Humankind now viewed existence from higher levels of Awareness; each embraced the Sacred and recognized the precious value of creation. ASELF was aware that Universe is alive, ever changing, evolving and purposeful, the world to be, a place of empowerment, responsibility, harmony, and balance. She was told that a spiritual peoplehood blends morality, ethics, law, and love into

One; dignity and safety are returned to Universe. ASELF watched Transpersonal Humankind gracefully move in and out of the inner Realms of SPIRIT; she knew they dwelled in and out of space and time, as well. Humanity was now the One and the Many.

Psyche, Earthling, and Humankind's new consciousness enthusiastically spilled out into the world and ushered in a renaissance of art and music, a Science of Consciousness, a Spiritual Psychology, and a compassionate Sacred Science. Creativity exuberantly flowed from the inner realms into the world of physicality. Humanity spans heaven and earth and co-creates Universe. And so it happened; an age of loving Consciousness unfolded. SPIRIT, SOUL, and Universe were ecstatic and jubilant, their noble dreams fulfilled.

Dream 19—Crossroads

ASELF's last dream presented her with a review of all she had previously learned. She had watched the genesis of planet and Universe, and had witnessed the emergence of earthly consciousness and its evolution into multitudinous life forms. ASELF witnessed the cycle of creation and annihilation endlessly unfolding throughout Universe as Consciousness pulsates in and out of existence with each breath. She recognized that the Cosmic Dance of Shiva is the dance of birth and death; it is energy embracing matter and matter embracing energy. ASELF watched living systems fall apart and magically come together again in more sophisticated and innovative ways. She now understood death to be as necessary to life as is birth; she understood the cosmic Law of old blending into new.

Then ASELF's dream shifted its tone and solemnly revealed a humanity standing at a dangerous crossroads. She was told that humanity would soon decide which of two diverging pathways to travel. One path approaches a dangerous precipice and descends into chaos. She realized that those who travel upon this road will simply disappear. Present-day, "Ordinary" humankind was rapidly approaching an evolutionary dead-end. At the end of this road, death and extinction await the arrival of Psyche, Earthling, and Humankind; she knew none would escape. (See Table 10.)

An alarmed and apprehensive ASELF awakened from her dream. She hoped humanity would survive; she wondered which path We would choose. ASELF knew that sages and mystics of every age traveled another path; theirs was the journey of spiritual awakening and embracing ASOUL. Travelers upon this second pathway yearned

Table 10
Evolution of Psyche, Earthling, and Humankind

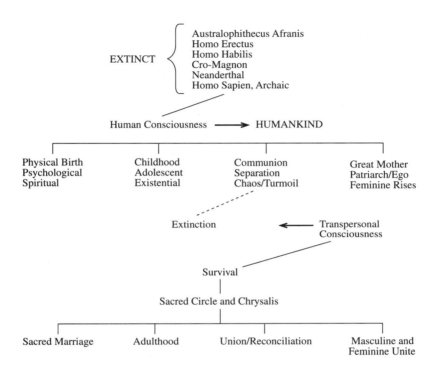

for the gifts of Transpersonal consciousness and inner peace. An astonished ASELF recognized the second road to be none other than the Path of the Heart and the spiral journey through the Realms of SPIRIT.

Once upon a time, the gift of Human consciousness had transformed Homo Sapien into humankind. Once upon a second time, humankind was being invited to accept the gift of Transpersonal consciousness and to become the sacred vessels for a radically new Specieshood. ASELF knew that the gift of expanded Consciousness would transform Psyche, Earthling, and Humankind, and allow all three to evolve light years beyond their present understanding of Self and Universe. ASELF hoped they would choose the Path of the Heart. She knew this second path would grant them eternal LIFE.

Destiny

ASELF spent many months contemplating and assimilating the wealth of information and ideas she had been given in her dreams. She realized that Evolution was swiftly propelling Psyche, Earthling, and Humankind onto still another stage of growth and transformation. Humanity was rapidly approaching a dangerous crossroads. Humankind must choose whether to change and grow, or to perish and die. ASELF cried out to earth's black, brown, red, white, and yellow peoples. She echoed the warnings of Higher Consciousness and beseeched humanity to take action before it was too late.

ASELF perceptively understood that two critical choices await humankind. First, humanity must immediately decide whether to perish or to survive. Earth can no longer sustain our irresponsible, destructive ways. Second, she knew that humanity must also decide whether or not to become the sacred vessels for Transpersonal humankind. She yearned for humanity to uncover its multi-faceted, multi-leveled, multi-dimensional Selfhood and affirm its partnership with the hallowed ground of ALL THAT IS.

ASELF acknowledged the eternal validity of Consciousness. She discerned that Consciousness and Universe were already preparing for the emergence of Transpersonal speciesship. She perceptively recognized that ALL THAT IS would *still* bestow the extraordinary gift of Transpersonal consciousness and create a new species of Awareness upon earth, *irrespective* of humanity's choice. She knew with certainty that SPIRIT's most fervent wish and dream to experience a Spiritual Renaissance upon Earth would prevail. Consciousness would continue to evolve and endure. (Refer to Table 11.)

> *"There is a fifth dimension beyond that which is known to man. It is a dimension as vast as space and as timeless as infinity... This is the dimension of imagination. It is an area we call 'The Twilight Zone.'"*[6]
>
> —*Rod Serling*

Table 11
Evolution of Old and New Paradigms

The Great Mother/Goddess: "First Paradigm"
Symbiosis—fusion
Oceanic Unity—Oneness
Unified Psyche
Obedience to Authority
Passive
Dependent—Vulnerable
Prepersonal/Pre-egoic
Protection, Sustenance, and Nurturance
Harmony
Sacred Time and Cycles

The Patriarch "Old Paradigm"
Separation, Individual, "I vs. Thou"
Either/Or, Black/White Thinking
Independence, Autonomy
Power, Control, and Domination
Deterministic, Predictable
Materialization
Technology and Specialization
Rational, Logical, Reality (Logos)
Linear Time Flow—Cause and Effect
Personal—Ego Consciousness/Shadow
Free Will—Choice
Responsible/Irresponsible
Multi-faceted

The Feminine Emerging "New Paradigm"
Receptive, Intuitive (Eros)
Free Will—Choice
Responsible/Accountable
Partnership—Relational—Reconciliation
Inner Empowerment
Commitment
Interdependence/"I vs. Thou"
    also become "We and Us"
Interconnected/Interrelated/Wholes
Spiritual Awakening
Gateway to Transpersonal—Trans-egoic
Simultaneous Time
Both/And Thinking
Uncertainty, Unpredictability
Evolving, Emergent
Harmonious
Honors and Respects Diversity/Duality
Multi-leveled

(table continued on next page)

Table 11
Evolution of Old and New Paradigms
(continued)

Sacred Marriage	**"Future Paradigm"**
Union	
Harmony and Balance	
Commitment, Intimacy	
Whole	
Mind, Body, and Spirit	
Non-Dual, Inclusive	
Sacred Time	
Co-Creator and Partnership	
Spiritual Insight	
Compassion	
Service—Global	
Multi-Dimensional	
Interdependence	
Diversity within Unity	
Direct Experience with Numinous	
Transcendental	
ONE	

Each organizing principle imparts a particular perception and understanding of Self, World, Universe, and Reality. Each world-view also determines our understanding of the arts, education, government, humanities, philosophy, psychology, the sciences, and spirituality. The Evolution of Psyche proceeds from pre-personal to personal to transpersonal/collective levels and culminates in transcendental awareness.

    ASELF had learned her mystical lessons well; she was granted the extraordinary gift of prophecy. She had learned the art of dreaming and to envision the future possibilities awaiting Psyche, Earthling, and Humankind. She endeavored to bring Light into a very dark, fragmented world. A visionary, mystical ASELF was stirring. She yearned for Psyche, Earthling, and Humankind to open their hearts and become the One We Are. She wished to assist and guide humankind on its sacred journey into a Brave New World.
    ASELF closed her eyes and thought about the world she wished for her children and for her children's children, and for the generations upon generations yet to be. ASELF entered the Twilight Zone and the World of Imagination. She began to yearn and to dream. And so it happened; she glimpsed the World of 2070 A.D.

Notes

1A. *Old Testament* Joel, III:1.
1B. *New Testament* Paul, Book of Acts, II:17.
2. Ojibuay Prayer.
3. Doug Boyd, *Rolling Thunder* (New York, Dell Publishing Company, 1974), p. 266.
4. T. S. Eliot, *Four Quartets* (Harcourt Brace & Company, 1943, 1971).
5. Andrea Cagan, *Caterpillar Dreams*, a poem, 1994.
6. Rod Serling, *Daily News TV Book*, Los Angeles, May 1994.

Selected Bibliography

Thomas Berry, *The Dream of the Earth* (San Francisco, Sierra Club Books, 1988).
Nigel Calder, *Timescale: An Atlas of the Fourth Dimension* (New York, Viking Books, 1983).
Rachel Carson, *Silent Spring* (Boston, Houghton Mifflin, 1987).
Riane Eisler, *The Chalice and the Blade: Our History, Our Future* (San Francisco, Harper & Row, 1987).
Jostein Gaarder, *Sophie's World* (New York, Farrar Straus and Giroux, 1994).
Marija Gimbutus, *The Civilization of the Goddess,* Ed. Joan Marler (San Francisco, Harper-San Francisco, 1991).
Al Gore, *Earth in the Balance: Ecology and the Human Spirit* (New York, Houghton Mifflin, 1992).
Stephen Jay Gould, *Dinosaur in a Haystack* (Random House, New York, 1996).
Hazel Henderson, *Paradigms in Progress: Life Beyond Economics* (Indianapolis, Knowledge Systems, Inc. 1991).
Rod Serling, in *Daily News TV Book*, May 15-21, Los Angeles, 1994.
Starhawk, *The Fifth Sacred Thing* (New York, Bantam Books, 1993).
David Suzuki and Peter Knudtson, *Wisdom of the Elders* (New York, Bantam Books, 1992).
Brian Swimme & Thomas Berry, *The Universe Story* (San Francisco, Harper-San Francisco, 1992).
Christopher Wills, *The Runaway Brain* (New York, Basic Books, 1993).
Robert Wright, *The Moral Animal* (New York, Panetheon Books, 1994).

2070 A.D.—The Wisdom Age

Section seven

The world of 2070 A.D. is a vision of a new era. It is a world filled with hope, promise, and opportunity. Its people have achieved a sustainable economy and a sustainable ecology. They live in balance and harmony with the cycles and rhythms of the cosmos, and respect all forms of life and consciousness.

The world of 2070 A.D. is founded upon the Sacred Marriage of masculine and feminine elements in Psyche, and a new spiritual covenant with Universe. The citizens of 2070 A.D. know the inner paths of the Psyche. They are equally at home in the worlds of matter and non-matter, for they have long recognized that they are One.

Balance in population, food supply, and energy resources ensure that every child is a wanted child and is properly cared for. Parenting is considered the most noble of vocations. The "Mothering Ones" are esteemed and cherished along with the children they nurture.

More mature and humane solutions to conflict and discord have been discovered and promoted. The world of 2070 A.D. knows no honor, glory, or profit through war.

Elders are valued for their abilities and experiences. Their presence and their gifts enrich the entire community. The Wisdom age is the culmination and gift of a life well-lived. Advancing age affords an opportunity to more fully embrace and explore the inner paths of transcendence.

Creativity and the willingness to grow are prized attributes in the land of 2070 A.D. People of all ages are encouraged to create, to risk, to fail, and to succeed. Essentially, they are free simply to *BE* whomever they are. Black, white, yellow, red, or brown, all citizens are equal and all are cherished. This is a world that honors and respects diversity. Value fulfillment is everyone's privilege; it is freely offered to all who dwell in this peaceful realm.

2070 A.D. inspires and encourages its citizens to become dreamers, seers, and prophets. They have learned that their dreams and future emanate from the invisible archetypal realms of the psyche, and beyond. The seeds of inspiration and wisdom are transmitted to 2070 A. D. through its visionaries and enlightened ones. No society can long endure without the arts, music, literature, and *visions* that reflect its highest potential. Archetypal 2070 A.D. dwells in the hearts and Souls of humankind. It symbolizes our promise and our birthright.

The enlightened, peaceful land of 2070 A.D. sharply contrasts with the shadowy, dark world we presently know. In our current world, we experience violence, despair, greed, hate, abuse, violation of human rights, and the desecration of Mother Earth. It seems unlikely, even futile, to believe that we could ever realize and attain such a reality.

We are like the people in Plato's Cave[1*], seeing only the shadows and hints of another reality. Like Plato's people, we are faced with an awesome choice. We can refuse to believe and change or, instead, we can embrace the Light and leave our Darkness behind. 2070 A.D. already exists within the realms of possibility. It is humankind's destiny and future. *Carpe diem.* Dare we seize the opportunity?

Humankind is approaching the next evolution of consciousness and perception that corresponds to Reconciliation and Union. It urges us beyond the limitations of gender, stereotyped roles, and the prejudice against race or creed. It imparts dignity and respect to all human beings. In the life cycle of ASELF, it represents the choices and responsibilities of conscious adulthood.

Recognition that Psyche continues to develop throughout one's entire lifespan implies that humankind might yet realize more periods of chaos and uncertainty as our species experiences middle-age, and as we accrue the wisdom and grace of an ancient people. Each stage in Psyche's cycle of life guides and directs our continuing sojourn upon Earth.

Humanity must successfully navigate and accomplish an inner reconciliation of masculine and feminine principles, in order to realize its fuller potential and to embrace the One and the Many. TRANS-PERSONAL Human will soon dwell upon the earth. The caterpillar is transforming itself into a butterfly. When the metamorphosis is complete, we will know the ONE WE ARE, consciously, for the *very* first time.

---

*"The dwellers in the den see only the shadows of puppets, which are themselves only imitations of really living things. That is, they see only the appearance of material things, not their true nature." Plato, The Republic, pp. 398–399.

PRAYER

ASELF awakened with a sigh,
She yearned to dwell in 2070 A.D.
She knew that you and I
would determine its reality.

ASELF beseeched one and all
to heed SPIRIT's Soulful call.
She encouraged us to awaken and become whole
Humankind would soon embrace ASOUL.

Notes

1. Plato, *The Republic* (Roslyn, Walter J. Black, Inc. 1942). pp. 398-399.

Appendix I

The Myth of Demeter and Persephone

The myth of Demeter and Persephone, from Homer's "Hymn to Demeter," reveals the fate of the feminine archetypal principle following her repression and banishment to the underworld. This event occurs after Patriarch slays the dragon (Great Goddess) and begins to govern Psyche; it accompanies Psyche's shift from pre-Egoic to Ego-centric consciousness. Zeus is Persephone's Father. Hence, Demeter and Persephone are Goddesses; Demeter is the Great Goddess of Nature, Woman, and Harmony; she often bestows the gift of immortality upon humankind. Worship of Demeter is at the core of the Eleusinian Mysteries.

As Persephone (also known as Kore) wanders away from her companions, she sees a beautiful Narcissus with a hundred blossoms. Unaware that this heavily scented flower has been placed there by Zeus (the Patriarch) as a form of entrapment, Persephone picks the flower. Instantly Earth opens, and Zeus's brother Hades, Lord of the Underworld, carries Persephone into the depths of the unconscious. The entire deed is witnessed by Helios the Sun. Zeus does not respond to Persephone's cries for help (we learn later that Zeus has planned the entire abduction).

Demeter hears her daughter's anguished cries for help, but cannot discover Persephone's whereabouts. For nine days, Demeter searches for her daughter and openly laments her absence. On the tenth day, the two goddesses, Hecate and Demeter, set out together to visit Helios the Sun. Helios tells them that Zeus, himself, conspired to abduct Persephone, who is to marry Hades. Helios urges Demeter to cease her grief and rejoice in her daughter's marriage to Hades.

But Demeter's grief increases and she continues her search for Persephone. Demeter, now disguised as an old woman beyond child-bearing age, arrives in the kingdom of Eleusis after long months of searching. She approaches several women drawing water from the Maiden's Well, and tells them that she is seeking work. They offer to take her to the palace to meet the King and Queen who need a nurse for their young son. Under Demeter's care, the prince grows stronger and more handsome each day. But one night, while the rest of the kingdom lay asleep, the Queen watches in horror as Demeter places her baby son upon the fire. In terror, the Queen snatches the baby away from Demeter. In a moment of rage, Demeter reveals her true identity, and informs the Queen that she has deprived her young son of immortality by snatching him away. Demeter then demands that a great Temple be built in her honor in Eleusis. Upon its completion, Demeter moves into the temple and continues to mourn for Persephone. During the next year, a withered and distraught Demeter, Goddess of Nature and Agriculture, punitively withhold's earth's crops and produce. As earth's bounty begins to wither and die, the Gods fear humankind will soon perish. They implore Demeter to restore fertility to earth. Demeter refuses; she wants her daughter back. Finally, Zeus requests that his brother Hades allow Persephone to return to her mother. Hades agrees and, as a parting gift, he offers Persephone a pomegranate seed.

Until now, Persephone has refrained from partaking of any food or sustenance in the underworld. Essentially, she has refused to be nourished and influenced by the unconscious. If Persephone had not eaten the seed, she would have continued her passive dependency upon Demeter. Only in this undifferentiated state could Persephone remain forever innocent. Yet in her joy, Persephone willingly eats the seed of the pomegranate, an action which compels her to return again and again to the world of darkness. Upon Persephone's return, Demeter restores earth's bounty and sustenance to humankind.

Significance

The myth of Demeter tell us that it is Patriarch (active principle) who slays the dragon, not the Feminine (passive principle). It is Patriarch who urges humankind toward autonomy, independence, separation, and specialization; a prerequisite for later emergence of the Feminine. And it is Patriarchal Hades who offers Persephone the pomegranate seed as a farewell gift. We learn that acceptance and ingestion of this seed prevents Persephone's regression to passive identification with Mother, although she must return to the underworld one-third of each year. We realize that Persephone has come of age. She not only chooses to eat the pomegranate seed, she also accepts responsibility for her actions. Demeter intuits that the daughter who returns to her is not the same one who was abducted. Demeter recognizes that her daughter has willingly eaten the bitter fruit of awareness in order to heal the split in Psyche. It is indeed a moment of great triumph as well as joy. Both Demeter and Persephone have passed through the shadow of darkness and emerged whole and renewed.

"Any breakthrough of new consciousness, though it may have been maturing for months or years out of sight, comes through a building up of tension which reaches the breaking point. If a man or woman stands firm with courage, the breakdown becomes a breakthrough into a surge of new life. If he cannot stand it and settles for an evasion, then he will regress into neurosis."

—Luke

References

Funk and Wagnalls, *Standard Dictionary of Folklore, Mythology and Legend* Maria Leach and Jerome Fried (Eds.), Harper and Row, San Francisco, 1972, pp. 306–307.

Luke, Helen M. *The Way of Woman: Awakening the Perennial Feminine* (Doubleday, New York, 1995), p. 108.

Appendix II

ASELF had a revelation: The different eras of Human *Civilization* (Communion, Separation, Chaos, and Uncertainty) corresponded to *Psyche's Evolution* through the *Spectrum of Consciousness* (Pre-Egoic, Ego, Existential, Transpersonal) as well as paralleled the descent and *evolution of Consciousness* through the *Four Worlds* of Origination, Creation, Formation, and Manifestation, and the *Great Chain of Being.* She discerned a remarkable correspondence between the *psychic principles* (Goddess/Mother, Patriarch/Masculine, Feminine, Sacred Marriage); they were describing One reality. ASELF experienced a magical moment of insight. She had discovered science, psychology, and spiritual traditions revealed the same Universe. Only their positions, levels, interest, tools, and perspectives varied. Amazingly, all disciplines and paradigms revealed the same Evolutionary thrust of Universe toward greater and greater expression, creativity, and community. She watched the many different pieces of a puzzle fit together to create a larger whole. ASELF wanted to know more.

Appendix III

Freud viewed the Oedipus complex as the *nuclear complex* of all psychoneuroses and, in his view, a variety of neurotic fixations, sexual aberrations, and debilitating guilt feelings were theoretically traceable to an "unresolved" Oedipus complex. Interestingly, the theoretical genesis of the complex—which derives its name from the mythical figure Oedipus, the hero of two of Sophocles' tragedies who unknowingly killed his father and married his mother—was Freud's own self-analysis, carried out after his father's death. At first, *Oedipus* referred to only the male complex, *Electra* being used for the female. Electra's sins were different than Oedipus'. Rather than directly murdering her mother, she urged her brother to do it. A different interpretation is presented in this book.

Glossary

Absolute time: Newton's view of time, in which time flows at the same rate throughout the universe and people at different locations experience the same "now."

AEther: A hypothetical medium formerly believed to fill all space and support the passage of light and other electromagnetic waves.

ALL THAT IS: A metaphysical term meaning the vast and infinite psychological realm whose energy is within and behind all formations. It is alive within the least of Itself. All of Its creations are endowed with Its own abilities, and It is infinitely becoming, never complete.

Amino acids: The molecular building blocks of proteins.

Anthropic principle: Idea that our presence in the universe puts constraints on its properties. More extreme versions of this principle border on the claim that the universe was designed for the benefit of humankind.

Anti-particle: The counterpart of an elementary particle with identical mass and spin and opposite electric charge and magnetic properties. When a particle and its anti-particle collide, both are destroyed, with the equivalent energy surviving as radiation. Some electrically neutral particles are their own anti-particles. Fortunately, anti-particles are usually seen only in laboratories; the rare exception is in the debris of a cosmic ray shower.

Archetype: A predisposition toward certain patterns of psychological performance (human and animal) that are passed down from our ancestral past and are linked to instinct. In Platonic terms, an archetype is the underlying form or idea of a thing. According to Jung, archetypes are forms without content in the collective unconscious, which become activated as psychic force.

Atman (Sanskrit): The primal source and ultimate goal of all, the one, divine reality, which pervades the manifold world of things and lives and minds. It is infinite and spaceless and timeless and yet the source for space and time.

Atman-project: Ken Wilber's term for the ascent of the human spirit up the spectrum of consciousness to fulfill its destiny and return to its true home, Atman.

Atom: A unit of matter consisting of a positively charged nucleus orbited by negatively charged electrons. The nucleus is made of protons and neutrons.

Attachment: Generally, a binding affection, an emotional tie between people. The *usual connotation* implies that this emotional relationship is infused with dependency; the persons rely on each other for emotional satisfaction. In *developmental psychology,* an emotional bond formed between an infant and one or more adults such that the infant will: (a) approach them especially in periods of distress; (b) show no fear of them, particularly during the stage when strangers evoke anxiety; (c) be highly receptive to being cared for by them and; (d) display anxiety if separated from them.

Autonomy: Independence.

Baqa: Affirmation of truth and love, process of becoming an "I."

Bardos: Buddhists describe six states of consciousness including waking, dreaming, meditation, dying, after-death, and re-birth.

Bifurcation: A point at which there are two distinct choices open to a system; similar to a fork at which a path divides into two. Beyond this critical point, the properties of a system can change abruptly.

Big Bang: A widely espoused cosmological theory according to which some 15 billion years ago all the matter and energy in the universe was born in a cataclysmic explosion. Since then the universe has expanded and cooled to its present state.

Black body: An object capable of efficiently absorbing all frequencies of electromagnetic radiation and emitting all frequencies when brought to incandescence. A true black body is a perfect absorber and radiator. Physicist Max Planck discovered the quanta while studying theoretical problems connected with black-body radiation.

Black hole: An object that is so dense that nothing, not even light, can escape from it except by quantum mechanical means. It is characterized by only three properties: mass, charge, and angular momentum.

Bonding: 1. Generally, the forming of a relationship. More specifically, that between the mother and her newborn. Some use the term as a synonym for *attachment*. 2. Others distinguish it as a separate process that occurs during the first few hours after the birth of the infant.

Buddha: A fully enlightened being; one who has overcome all negativities and completed all good qualities.

Causality: The doctrine that everything has a cause which is antecedent in time.

Cell: The ultimate component of all living organisms, the smallest unit that can function independently.

Chakra: Sanskrit word for the 7 psychic energy centers and focus of consciousness located along the spinal cord. One relates to life situations from the particular perspective and awareness of each chakra. Native Americans recognize 5 energy centers and Jewish Kabbalists describe up to 11.

Chaos: Term used to describe unpredictable and apparently random behavior in dynamical systems.

Chrysalis: Protective chamber that houses a caterpillar during its metamorphosis and transformation into a butterfly.

Circumincession: Unity in Divinity—with one aspect indwelling within the others.

Classical mechanics: The mechanics formulated by Isaac Newton.

Clear Light: There is no deity in Buddhism. Clear Light is the spark of eternal life and sacredness within all things. It is similar to the western understanding of Soul.

Closed system: One that exchanges energy but not matter with its surroundings.

Closed universe: A universe with sufficient matter to recollapse to a singularity.

Collective Unconscious: The portion of the psyche whose unconscious contents are hereditary and belong to all humankind.

Conciliation: 1. To overcome the distrust or hostility of; placate; win over. 2. To win or gain. 3. To make compatible; reconcile.

Consciousness: In this text, a particular quality of being, an infinite spectrum of layers which are not separate but mutually interpenetrated. Bohm refers to these layers as implicate orders, Wilber as levels, and others as frameworks or dimensions. All would agree that each individual layer is an aspect of an underlying whole and all layers are accessible with the proper focus. We are all on a journey to widen our focus.

Conservation of energy: An alternative statement of the First Law of Thermodynamics which says that energy can neither be created nor destroyed during any process.

Contemplation: A different kind of concentration that is difficult to describe in words. It comes close to Zen meditation. By way of introspection and conscious receptivity, contemplation observes the chosen subject. The purpose of contemplation is expansion; it is to lose oneself in order to find oneself in an altered state of consciousness. It happens when one ceases to exist and starts to be.

Copenhagen interpretation: The orthodox interpretation of quantum mechanics due to Bohr, Heisenberg, and their followers in which no reality can be ascribed to the microscopic world.

Cosmology: The study of the origin and nature of the universe.

Cyberspace: An artificial, mental environment formed through the interaction of human consciousness with a computer generated Virtual Reality.

Dependency: 1. In social and personality psychology, a condition holding between two or more persons in which one relies upon the other(s) for emotional, economic, or other support. 2. A characteristic of an individual in such a dependent state; a lack of self-reliance.

Dharma spiritual teachings: The doctrine of the Buddha; universal law.

Determinism: The doctrine that events are completely determined by previous causes rather than being affected by free will or random factors.

Deep structure: The defining form of each level of consciousness in Ken Wilber's holoarchy. Each level is divided into a deep structure and surface structure. The deep structure contains all the potentials and limitations of a given level. The surface structure is a particular form of the deep structure; within the potentials and limitations of the deep structure, the surface structure can assume any pattern.

Dissipative structure: An organized state of matter arising beyond the first bifurcation point when a system is maintained far from thermodynamic equilibrium.

Dissociation: Disturbance or alteration in the normal integrative functions of identity, memory, or consciousness; diminished response to the environment; an altered state of consciousness, detachment, self-observation from a distance.

DNA (Deoxyribonucleic acid): A very large nucleic acid molecule carrying the genetic blueprint for the design and assembly of proteins, the basic building blocks of life.

Dream Time: Timeless, formless, spiritual realm of Australian Aborigine, in which creation occurred and is still occurring.
Dukkha: Buddhist term for suffering.
Dynamical systems: General term for systems whose properties change with time. Dynamical systems can be divided into two kinds, conservative and dissipative. In the former the time evolution is reversible, in the latter it is irreversible.
Dynamics: The science of matter in motion.
Earthling: An individual member of Humankind. ASELF is an Earthling.
Ecology: Broadly, the study of the relationship between organisms and their environments. The discipline is concerned with the complex interrelationships between the various plants and animals with each other, and with the physical environment in which they live. There are several subdisciplines in the field, dealing with plant ecology, animal ecology, and human ecology. In recent years, the term has become increasingly popular owing to the recognition that virtually every act of a particular organism or any modification of the physical environment has a widespread impact on ecological structure.
Ecozoic: The emerging period of life following the Cenozoic and characterized, at a basic level, by its mutually enhancing human-earth relatedness.
Ego: a complex of ideas which constitutes the center of one's field of consciousness, and which appears to possess a high degree of continuity and relatedness.
Electron: An elementary particle of negative electric charge which orbits the nucleus of an atom and carries electrical current.
Enlightenment: The state of being a Buddha, when all duality is transcended into absolute unity; the eradication of all negative states of mind and the accumulation of all positive qualities.
Entelechy: The realization or actuality as opposed to a potentiality. A vital agent or force directing growth and life.
Entropy: In thermodynamics, a measure of the capacity of an isolated macroscopic system for change. Can be negative or positive entropy.
Epilogue: Post-script, Afterword, Coda.
Epiphenomenon: A phenomenon that arises from the organization of a particular system but is not present in its constituent parts.
Epistemology: A branch of philosophy that investigates the origin, nature, methods, and limits of human knowledge.
Equivalence principle: The equivalence between acceleration and the force of gravity.
Eukaryote: A cell with a membrane-bound nucleus, membrane-bound organelles, and chromosomes in which DNA is combined with special proteins.
Events: Points in four-dimensional space-time, which do not exhibit extension or duration. In the *Newtonian* view, the universe consists of things, whereas in the *Einsteinian* view, the universe consists of events, e.g., the interaction of subatomic particles.
Evolution: General term for the unfolding of behavior with the passage of time. Development or growth that produces higher and higher wholes; a

movement from the lower to the higher. In biology, the Darwinian theory according to which higher forms of life have arisen out of lower forms with the passage of time.

Explicate Order: Bohm's term for the domain referred to by Cartesian coordinates (locality in space-time). It displays the separateness and independence of fundamental constituents and is manifest or visible (directly or with instruments). It is secondary to the implicate order (see below), which unfolds to create the explicate order and enfolds to give guidance to itself.

Faña: Dissolve present status and reintegrate at a higher level. It is the removal of the "I."

Field: An area of space characterized by a physical property that is normally invisible and intangible but under certain circumstances can interact with matter. Classically, the action between two material objects separated in space is described in the language of fields. In *quantum field theory,* the interaction is viewed as an exchange of so-called messenger particles, which conveys a particular force. Examples are the photon (electromagnetic) and graviton (gravitational).

First Law of Thermodynamics: See "Conservation of energy."

Fractal geometry: The geometry used to describe irregular patterns. Fractals display the characteristic feature of self-similarity, an unending series of motifs within motifs repeated at all length scales.

Frame of reference: Technical term for the point of view of an observer in relativity, according to which all such frames are of equal status. A set of coordinate axes by which the position or location of an object may be specified.

Frameworks: Seth's term for interpenetrating levels of action or existence; analogous to the spectrum of consciousness. Our outer reality is Framework 1, and our inner reality and the creative source from which we form events is Framework 2. We draw from Framework 2 by focusing our attention properly. All That Is contains an infinite number of frameworks.

Gene: A unit of heredity composed of DNA, responsible for passing on specific characteristics from parents to offspring.

Genome: The entire genetic material of an organism.

Gnosis: Combining forms from the Greek for *knowledge* and used widely to denote knowing, cognition, recognition, etc.

God Particle: Also known as the Higgs particle, a.k.a. the Higgs boson, a.k.a. the Higgs scalar boson.

Godhead: essential being of God, Supreme Being, Deity.

Gravitation: The universal force of attraction between all forms of matter.

Ground unconscious: Ken Wilber's term for all the deep structures of all levels of consciousness existing in potential form, which eventually emerge as a particular surface structure within the constraints of the deep structure of that level.

Hidden variables: Variables that would allow physicists to define the quantum state more precisely than present quantum theory allows. If such variables exist, they probably will be discovered on the subquantum level where present quantum theory may be incomplete. Hidden variables would select among the given probabilities of the wave function.

Hierarchy: Ranking and ordering of levels or fields.

Hilbert space: The abstract space used by physicists to describe quantum mechanical states. A single point in Hilbert space represents the entire quantum system. Unlike the classical concept of phase space, Hilbert space is a *complex* vector space. The wave function is represented by a vector (state) composed of a superposition of states, each corresponding to a possible result of a measurement.

Holoarchy: Non-linear, non-sequential, asymetrical, interdependent, and interactive self-organizing systems, with each successive level containing the previous developmental level; fields within fields.

Hologram: A photographic record created by laser light directly reflected from an object and laser light going directly to the plate to form a complex interference pattern. When the plate is illuminated with coherent light, the wavefront produced creates a three-dimensional image of the original object. In this way, the whole image can be recreated from each portion of the record.

Holomovement: In David Bohm's *terminology,* the total ground of that which is manifest. The manifest is embedded in the holomovement, which exhibits a basic movement of unfolding and enfolding.

Holon: a whole at one stage, yet part of a wider holon at the next.

Homo Erectus: Human species dating from 1.5 million earth years ago to 300,000 earth years ago.

Homo Habilis: First human species, appearing around 2.6 million earth years ago in Africa.

Homo Sapiens: The only surviving species of humans, dating back some 300,000 earth years ago.

Hyperspace: Space that is defined by more than three dimensions. As in Euclidean space, all axes are normal to each other.

Id: In the Freudian tripartite model of mind, the primitive, animalistic, instinctual element, the pit of roiling, libidinous energy demanding immediate satisfaction. It is regarded as the deepest component of the psyche, the true unconscious. Entirely self-contained and isolated from the world about it, it is bent on achieving its own aims. The sole governing device here is the *pleasure principle,* the id being represented as the ultimate hedonist. The task of restraining this single-minded entity is a major part of the ego's function.

Identity: The earliest awareness of a sense of being, of entity—It is not a sense of *who* I am but *that* I am; as such, this is the earliest step in the process of the unfolding of individuality.

Imaginal Realm: The world of the *Shaman,* outside space-time. May be an overworld or underworld.

Implicate Order: The basic order, according to *Bohm,* from which our four-dimensional world springs. It is multidimensional, and its connections are independent of space and time. The implicate order is identified with the wave function in quantum theory. See "Explicate Order."

Glossary

Individuation: The process of integration of the personality; the quest for meaning.

Ineffable: 1. Incapable of being expressed or described; inexpressible; unspeakable. 2. Not to be spoken; unutterable.

Inner Child: Hidden, repressed, split off and disowned aspects of Self which reside deep within the personal unconscious.

Interference pattern: The pattern produced when two waves overlap. The relationship between the overlapping waves dictates the pattern: a crest will add to another to make a larger crest, while a crest and a trough will cancel.

Introjection: 1. Generally, the process by which aspects of the external world are absorbed into or incorporated within the Self, the internal representation then taking over the psychological functions of the external objects. 2. Specifically, in *psychoanalysis,* that process when the parent figures are the external objects and the *introjects* (as they are called) are the values of the parents; the process here leads to the formation of the superego.

Involution: In Ken Wilber's *terminology,* the movement of Atman down through the spectrum of consciousness to create the manifest world; the process by which the higher levels of being are involved with lower levels. See "Microgeny."

Irreversibility: The one-way time evolution of a system, giving rise to an arrow of time.

Irreversibility paradox: The paradox arising from the fact that macroscopic systems are irreversible, whereas microscopic ones are reversible.

Kabbalah: To receive—from the Jewish mystical tradition.

Karma: The law of cause and effect, of action and doing; the enduring pattern of behavior and consequences of our thoughts, speech, and actions in this and future lives.

Koshas: Hindus describe Five Koshas or sheath-like structures surrounding each individual. Each kosha corresponds to a particular level or grade of awareness.

Kundalini: Personal, feminine form of fundamental life force, Prana, which pervades all of existence. Also known as the Holy Spirit, Ch'i, and Ki.

Laser: Acronym for Light Amplification by Stimulated Emission of Radiation. Laser light is highly coherent and has extremely large photon densities.

Least Action Pathway: Are paths of least resistance. Through repetition and positive reinforcement, it becomes easier and more efficient for energy and experience to travel this pathway.

Light: In this book, light refers to the entire electromagnetic spectrum. It can be interpreted as a wave, as in *classical* theory, or as a stream of quanta or photons, as in *quantum* theory. The frequency of the wave is related to the energy of the photon by *Planck's* constant. The velocity of light in free space is constant regardless of the motion of the observer. *Bohm* sees the implicate order as an ocean of energy, or light. Matter is a condensation of this energy; it arises when light rays reflect back and forth to form a pattern. Without reflection, there is simply pure light. Matter carries with it time and space, but pure light does not.

Macroscopic Systems: Systems of everyday dimensions such as billiard balls, tables, people, and so on, for which thermo-dynamics and Newtonian mechanics are usually applicable.

Mass: Either the property of matter through which the force of gravity acts (gravitational mass) or the resistance of a body to acceleration (inertial mass). The two are identical by *Einstein's equivalence principle*.

Matter: According to *Bohm, Wilber,* and *Seth,* matter is condensed consciousness. The unfolding and enfolding process creates successive localized manifestations that appear to our senses and instruments as physical form.

Maya: *Hindu* and *Buddhist* term for the relative world of illusion which many consider as the only reality. It is the world of measurement and physicality.

Mechanism: The philosophy that envisions the universe as composed of basic entities that follow an absolute set of quantitative laws. The mechanistic view is that if these laws were known and the intellect were large enough, all future events could be calculated and predicted with complete precision. This idea emerged from an extrapolation of Newtonian mechanics to all possible domains of knowledge.

Meditation: A continuous and intense flow of clear thoughts around the object of concentration in its narrowed viewing range. The result will be an impersonal and deeper metaphysical awareness.

Meta: Beyond, transcending, higher.

Metamorphosis: 1. An abrupt transition in form and structure as occurs in insects (egg, larva, pupa, adult). 2. A transformation in personality.

Metaphor: A word, phrase, or description applied to something that it does not literally designate, thereby suggesting comparison or analogy. For example, the Newtonian universe is often referred to as a machine.

Microgeny: In Ken *Wilber's* interpretation of the *Perennial Philosophy,* the moment-to-moment involution of the spectrum of consciousness. See "Involution."

Microscopic Systems: Systems of atomic and molecular dimensions ruled by quantum mechanics.

Molecule: The smallest unit of a chemical compound which still possesses the properties of the original substance.

Moment point: *Seth's* term for the *present* moment, the point of interaction between all existences and realities, through which all probabilities flow. According to Seth, the past, present, and future exist together and can be experienced in a moment. This is analogous to *Minkowski's space-time* in which all points of space and all points of time exist en bloc. The moving present we perceive as time is, in Seth's concept, a projection from a higher-dimensional timeless order.

Morphogenesis: The evolution of form in animals and plants.

Muon: Also known as a mu meson. Particles that are heavy relatives of electrons.

Mutation: A change in the genetic material of an organism.
Mystic: One who is capable of transcending the physical realm of space and time through a state of consciousness that extends beyond the restrictions of the intellect to unity with the absolute.
Mystical: 1. Mystic; occult. 2. Of or pertaining to mystics or mysticism. 3. Spiritually symbolic. 4. Rare. Obscure in meaning; mysterious.
Nadis: Points of entry for Kundalini energy into Chakra system. Also correspond to acupuncture meridians.
Nagual: In Central-American traditions, refers to the mysterious, unmanifest unknown.
Namasté: Sanskrit word for "the Divinity within me honors and recognizes the divinity within you."
Neutrino: An uncharged elementary particle generally believed to have no mass. Its anti-particle is an anti-neutrino.
Non-linear systems: Behavior typical of many real systems, meaning in a qualitative sense of getting more than you bargained for, unlike linear systems, which produce no surprises. Dissipative non-linear dynamical systems are capable of exhibiting self-organization and chaos.
Non-locality: Action at a distance. When something happens at one point there are immediate, simultaneous consequences immediately and over the whole of space, unrestricted by the velocity of light. Information is sent across space and time without passing through the space in between. Non-locality implies another dimension or level of reality beyond space-time, one in which everything is interconnected and interrelated with everything else.
Numinous: Refers to the spiritual, supernatural, and the mysterious which surpasses comprehension and understanding. Ineffable.
Oedipus Complex: A group or collection of unconscious wishes, feelings, and ideas focusing on the desire to "possess" the opposite-sexed parent and "eliminate" the same-sexed parent. In the traditional *Freudian* view, the complex is seen as emerging during the Oedipal stage, which corresponds roughly to the ages 3 to 5, and is characterized as a universal component of development irrespective of culture. The complex is assumed to become partly resolved, within this *classical* view, through the child making an appropriate identification with the same-sexed parent, with full resolution theoretically achieved when the opposite-sexed parent is "rediscovered" in a mature, adult sexual object.
Ontogeny: From the roots of the word, the origin and, by extension, the development of an *individual* organism. *Development and developmental* are generally preferred in discussions of child psychology.
Open system: One that can exchange energy *and* matter with its surroundings.
Open universe: A universe with insufficient matter to collapse to a Big Crunch.
Paradigm: In *science,* a conceptual framework, endorsed by the large majority of the scientific community, that presents problems to solve and defines boundaries within which solutions are sought. The term is often applied to the *societal realm,* in which it refers to the concepts and values of a community that shape its perception of reality.

Patripsych: The internal constellation of patriarchal patterns, attitudes, ideas, and feelings that develop in relation to authority and control. It is part of the *cultural* unconscious.

Perennial Philosophy: The term popularized by Aldous Huxley to refer to the consensus of mystics from many ages and cultures that a transcendental unity is the reality of all things.

Persona: From the Latin, meaning person. In *classical Roman theater*, it was a mask which the actor wore expressing the role played. By extension of this notion, *Jung* used the term in his early formulations to refer to the role a person takes on by virtue of the pressures of society. It alludes to the role that society expects one to play in life and not necessarily the one played at a deep psychological level. The persona is *public*, the face presented to others.

Personality: A gestalt made up of many strands of consciousness that collectively creates a sense of Self. This is the "You, I, and me" we call "Self."

Personal Unconscious: Comprises the personal experiences that have been repressed and forgotten. Consists of preconscious, sub-conscious, and unconscious levels.

Photon: A quantum "particle" carrying the energy in electro-magnetic radiation. A unit of light.

Photosynthesis: A biological process which converts sunlight, water, carbon dioxide, and minerals into oxygen and energy rich nutrients.

Phylogeny: The origin and, by extension, the evolution of a *species* or other form of animal or plant; its evolutionary history. Contrast with *ontogeny*, which refers to the origin and development of an *individual* organism.

Pig parent: An internalized form of cultural oppression.

Pion: Also known as a pi meson. An unstable particle, responsible for binding protons and neutrons within the nucleus.

Plenum: Space that is completely filled with matter/energy; the opposite of vacuum.

Positivism: The doctrine that we can have no knowledge other than that provided directly by our senses; according to this philosophy, it is pointless to talk about atoms and molecules.

Positron: The positively charged anti-particle of an electron.

Probability distribution function: A mathematical function used in *classical mechanics* to work out how probable it is that a system occupies a given state in phase space.

Projection: A trait, attitude, feeling, or bit of behavior which actually belongs to your own personality but is not experienced as such; instead, it is attributed to objects or persons in the environment and then experienced as directed toward you by them, instead of the other way around. The projector, unaware, for instance, that he is rejecting others, believes that they are rejecting him; or, unaware of his tendencies to approach others sexually, feels that they make sexual approaches to him.

Prokaryotes: Single-cell organisms without nuclei (bacteria) that were the first life forms of earth. They predominated for approximately 2 billion earth years and still compose a large segment of the life community.

Psyche: The Mind, The SELF; The totality of all psychic processes, conscious and unconscious. In this book, it is the part of consciousness emanating from Soul, and which mediates between the Soul and other (physical and non-physical) levels of reality, or portions of its SELFHOOD.

Quantum: (pl. quanta) A discrete unit of energy in quantum theory.

Quantum field theory: The theory that results from the application of *quantum mechanics* to the behavior of a field. *Classically,* all fields in physics were treated as continuous, extending to all points of space. However, when quantum mechanics was applied to the field, it became quantized and exhibited definite energy states. This allowed for a particle interpretation and made the wave-particle duality somewhat more understandable.

Quantum mechanics: The mechanics that rules the microscopic world, where energy changes occur in abrupt quantum jumps.

Quantum potential: According to David Bohm, a potential that acts on a subatomic particle in addition to normal classical potentials. When two particles are separated in space, unconnected by any classical potentials, they are still correlated through the quantum potential, which acts instantaneously. Instantaneous action violates the spirit of relativity, at the least. It is through the quantum potential that Bohm arrives at his concept of wholeness wherein the two particles are not seen as separate entities but as projection from a higher-dimensional domain.

Quark: Quarks come in six flavors: Bottom (Beauty), Top (Truth), Up, Down, Charm, and Strange. Quarks combine with leptons to form all existing matter. Only the Up and the Down quarks are found in today's universe.

Quasar: Quasi-stellar radio sources—heavenly bodies first detected by their radio emissions. Believed to be the most distant objects in the universe.

Reductionism: A doctrine according to which complex phenomena can be explained in terms of something simpler. In particular, *atomistic reductionism* contends that macroscopic phenomena can be explained in terms of the properties of atoms and molecules.

Relativity: *Einstein's* theories of relativity, an extension of *Newtonian* physics, deal with the concepts of space, time, and matter. *Special* relativity starts from the premise that the laws of physics are the same for observers moving at constant speeds relative to one another. *General* relativity is based on the idea that the laws of physics should be the same for all observers, regardless of how they are moving relative to one another. In the *general* theory, gravity is explained as the curvature of space-time.

Reincarnation: Belief that an aspect of Self and/or consciousness survives physical death and is later reborn in another body existing in a future time period. This new person experiences itSelf as a completely different personality. This cycle of birth, death, and re-birth ends when the individual balances his/her Karmic debt to ALL THAT IS and attains insight, compassion, and enlightenment.

Responsibility: Acknowledgment and recognition of the choices and actions one has made during his/her lifetime. Acceptance of one's feelings, understanding, perceptions, and experiences as his/her own, without blame, anger, apology, or qualification.

RNA: (ribonucleic acid) The genetic material used to translate DNA into proteins. In some organisms, it can also be the principal genetic material.

Revelation: Disclosure, unmasking, discovering, uncovering, proclamation.

Sacred Marriage: Developmental stage in which masculine and feminine principles of the psyche unite and cooperate, creating a larger gestalt of awareness and wholeness.

Schrodinger's equation: An equation that describes the time evolution of a quantum state as represented by a *quantum wave function*. It is a complex differential equation which is completely deterministic but when a measurement is made, the quantum state is abruptly changed to one of a set of new possible states for which the probability of occurrence can be computed from the wave function.

Second Law of Thermodynamics: The law stating that, during an irreversible process, entropy always increases. The future state of any isolated system has higher entropy than its present or past states.

Seer: 1. A person who sees; observer. 2. A person who prophesies future events; prophet. 3. A person who is endowed with profound moral and spiritual insight or knowledge; a wise man or sage who possesses intuitive powers.

Self: The center of Personality; multi-faceted identity including conscious, pre-conscious, sub-conscious, and unconscious aspects.

SELF: The more inclusive, whole SELF, encompassing Personal, Collective, Transpersonal, and Transcendental levels as well as hidden dimensions. It is our *Soul* essence, the core of our being.

Self-organization: The emergence of structural organization. It occurs within dissipative non-linear dynamical systems.

Separation: Intrapsychic achievement of a sense of separateness from mother and through that, from the world at large. This very sense of separateness is what the psychotic child is unable to achieve. A sense of separateness gradually leads to clear intrapsychic representations of the Self as distinguished from the representations of the object world, and enables the child's sense of being a separate person.

Separation anxiety: 1. In *psychoanalysis,* the hypothesized anxiety on the part of the infant or child concerning possible loss of the mother object. 2. By extension, anxiety over the possible loss of any other person or object upon whom one has become dependent.

Separation-Individuation Phase: Margaret Mahler's term for the child's awareness of its discrete identity, separateness, and individuality apart from the mother. In her theory, this process follows the *symbiotic stage,* when the mutually reinforcing relationship between mother and child is dominant.

Shadow: 1. In *Jung's* approach, one of the archetypes; a complex of undeveloped feelings, ideas, desires, and the like—the "animal" instincts passed along through evolution to Homo sapiens from lower, more primitive forms that represent the negative side of personality; the "alter ego." 2. To follow.

Shaman: A medicine man; one, acting as both priest and doctor, who works with the supernatural.

Simultaneous Time: Concept that all Time occurs simultaneously in a nonlinear realm. Past, Present, and Future occur in the Spacious Present.

Singularity: A region where the mathematics underpinning a theory becomes ill defined.

Sorcerer: A practitioner and intermediary between the natural, everyday world and an unseen world referred to as "The Second Attention." One who teaches the Art of Dreaming.

Soul: Eternal core of SELFHOOD and being. Lives many lives in many worlds. This is the essence of one's full identity. It is the God-like aspect of SELF.

Space-time: Four-dimensional synthesis of space and time occurring in relativity. Can be understood as projections from the space-time world.

Spectrum of consciousness: The hierarchy of increasing dimensions of consciousness, according to the *Perennial Philosophy*. The lowest level is the most dense and fragmentary (matter and energy), the highest level the most subtle and unitary (All That Is).

Spiral: A helix, formation, or form.

Spirituality: Being at one with the Universe, in tune with the infinite. Aware of God in all things, all events, and all circumstances.

Statistical mechanics: The discipline which attempts to express the properties of macroscopic systems in terms of their atomic and molecular constituents.

Superego: In the *Freudian* tripartite model of the psyche, the hypothetical entity associated with ethical and moral conduct and conceptualized as responsible for Self-imposed standards of behavior. The superego is frequently characterized as an internalized code or, more popularly, as a kind of *conscience,* punishing transgressions with feelings of guilt. In the *classical* psychoanalytic literature, the superego is assumed to develop in response to the punishments and rewards of significant persons (usually the parents), which results in the child becoming inculcated with the moral code of the community.

Super Implicate Order: A super-information field that makes the implicate order designated by the quantum wave function nonlinear and thereby organizes into relatively stable forms (explicate order). It is also a wave function or a super-wave function.

Symbiosis: Refers to an intrapsychic rather than a behavioral condition; it is thus an inferred state. This is a feature of primitive cognitive-affective life, wherein the differentiation between self and mother has not taken place, or where regression to that self-object undifferentiated state has occurred. Indeed this does not necessarily require the physical presence of the mother, but it may be based upon primitive images of oneness and/or disavowal of contradictory perceptions.

Synchronicity: Acausal connecting principle; a meaningful coincidence when an inner and outer event come together.

Tao: Chinese word for eternal way of the cosmic order.

Thermodynamics: The science of heat and work.

Time dilation: The effect of time slowing down, caused by increasing speed or the pull of gravity. In the latter case it is known as gravitational time dilation.

Time-symmetry: The property of Newtonian, Einsteinian, and quantum mechanics that both directions of time are equally permissible.

Tonal: In Central-American traditions, denotes the ordinary, physical world.

Transcendent: Surpassing, extraordinary, going beyond what is given or present in experience, hence beyond knowledge. Of God, being prior to and exalted above the Universe, and having being apart from it, beyond the natural and rational, the unknowable character of ultimate reality. Asserts the primacy of the spiritual and superindividual as against the material.

Transformation: Generally, modification or change in the form or structure of something. Clearly, a very rich conceptual notion and, not surprisingly, the term finds wide application in various specific domains. In *psychoanalysis,* the altering of a repressed impulse or emotion so that in its disguised form it can be "admitted" to consciousness.

Transpersonal Psychology: Transpersonal Psychology is concerned with the study of humanity's highest potential, and with the recognition, understanding, and realization of unitive, spiritual, and transcendent states of consciousness. Also called the Fourth force in psychology.

Uncertainty principle: A quantum mechanic principle stating that there is a fundamental limitation on the simultaneous measurement of pairs of quantities such as position and momentum.

Union: 1. The act of uniting two or more things; the state of being united; combination; conjunction. 2. Something formed by uniting two or more things; combination.

Vijnanas: Buddhists describe eight levels or states of Consciousness.

Virtual particle: A quantum particle that appears and disappears spontaneously and whose existence is of an extremely short duration. The *Heisenberg* uncertainty principle allows for this phenomenon, and as a result these short-lived particles are considered to be different from the more familiar observable particles.

Virtual Reality: Now called *Immersive Technology*. Produces synthetic computer simulations of both real and artificial mental environments, termed *cyberspace*. This technology enables the human mind to learn, perform, and render decisions well beyond ordinary human perception and ability.

Void: Nothingness. Also invented by *Democritus*. A place that atoms can move around in. Today's theorists have littered the void with a potpourri of virtual particles and other debris. Modern terms: the vacuum and, from time to time, the aether.

Glossary

W+, W-, Z⁰ family, and gluons: These are particles, but not matter particles like quarks and leptons. They transmit the electromagnetic, gravitational, weak, and strong forces, respectively. Only the graviton has not yet been detected.

Wave frequency: The number of complete cycles per unit time of a vibrating system.

Wave function: A mathematical description of the state of a quantum system; a solution of *Schrodinger's* wave equation.

Wavelength: The distance between two successive crests of one wave.

Wave packet: The configuration that results when a wave function is confined to a small region of space. This implies that the particle being described is confined to a fairly localized position.

Wave-particle duality: The twofold nature of the quantum objects, which can display wavelike and particle-like properties.

White hole: A hypothetical time-reversed black hole, is a source for all matter. Mathematically, it is possible that a black hole is connected to a white hole through a tunnel that may end up in another part of the universe or in a second universe.

Wholeness: Bohm's concept of an undivided, flowing movement, unbroken and all-encompassing but not complete and static. Division or analysis into an arrangement of objects and events can be appropriate for a limited description or display but is not the primary reality or ground.

Wormholes: Passages from one region of space-time to another.

Yang: Masculine, active, positive, and rational aspect of cosmic duality.

Yin: Feminine, passive, negative, and receptive aspect of cosmic duality.

Zero point energy: The energy possessed by atoms, molecules, or fields at absolute zero, the lowest achievable temperature.

Scientific Terms from

Coveney and Highfield, *The Arrow of Time* (New York, Fawcette Columbine, 1990), pp. 360-365.

Friedman, N. *Bridging Science and Spirit* (St. Louis, Living Lake Books, 1994).

Brian Swimme & Thomas Berry, *The Universe Story* (NewYork, Harper-Collins, 1994).

Peter Lafferty & Julian Rowe (Eds.) *The Dictionary of Science* (New York, Simon & Schuster, 1994).

Buddhist and Jungian Terms from

Moacanin, R., *Jung's Psychology and Tibetan Buddhism* (London, Wisdom Publications, 1986), 119-121.

Alan Watts, *Cloud-Hidden* (New York, First Vintage Books, 1974).

John White (Ed.) *Frontiers of Consciousness* (New York, Avon, 1974), p. 38.

Psychological terms from

The Penguin Dictionary of Psychology (London, Penguin Books, 1985).

Definition of Transpersonal Psychology from

La Joie and Shapiro, *Definitions of Transpersonal Psychology* in The *J. of Transpersonal Psychology,* Vol. 24, No. 1, 1992.

Additional Sources

Webster's Encyclopedic Unabridged Dictionary of the English Language (New York, Portland House, 1989).

Index

15 billion earth years, 148, 151–152, 180
13th Century, 117
16th Century, 118, 152
17th Century, 118
20th Century, 152, 155
21st Century, 146, 152, 157, 171
1.5 million earth years, 151, 184
2.6 million, 151–152, 184
5 million earth years, 151, 152, 184
40,000 earth years, 151, 153
200,000 earth years, 151
2070 A.D., 168, 171–174
9 orbiting planets, 149
4-dimensional, 13, 35, 25
A "WHERE," SOMEWHERE, 13
ABSOLUTE
   ALL THAT IS, 35, 43, 98, 109, 111–112, 114, 119, 121, 127–128, 144, 147, 161–163, 166, 179, 189, 191
   SOURCE, 12, 15, 27, 38, 42, 64, 86–87, 99, 106, 108–109, 112–113, 115–116, 122, 125, 127, 134, 137, 148
   SPIRIT, 2, 19, 61–64, 67, 86–87, 91, 93, 96–99, 101–109, 111–113, 115, 118–120, 124–125, 128–129, 131, 133, 136–137, 139–141, 144, 147–148, 151, 157–159, 162, 164–166, 168–169, 174, 185, 193
absolute or Superspace
   space, 2, 5, 8, 10–16, 18–20, 25–29, 31–32, 35, 37, 42, 90–91, 93, 106, 108–111, 117, 122–123, 125–128, 132–133, 146, 148, 151, 158, 160, 164, 166, 179, 183–189, 191, 193
academic, 75–78, 91, 145
accomplishment, 74, 127
act of Creation
   Big Bang, 10, 12–14, 20, 32, 180
   Genesis, 11, 60, 107–108, 121, 147–148, 164
actions, 6, 31, 77, 90, 101, 123, 127, 129, 131–132, 145, 176, 185, 190
actors, 123
adolescence
   Ego Consciousness, 55–56, 130, 154–155, 167
   Patriarch, 54, 58, 73, 89, 91–92, 101, 125, 154–156, 159, 167, 175–177
   Separation, 28, 31, 43, 53–56, 58, 71, 76–78, 83, 86, 90, 101, 125, 153–155, 159, 167, 176–177, 190
adulthood
   ASELF, 2, 70–94, 105, 110–112, 114, 126, 134–135, 138, 144–151, 153–166, 168, 173–174, 177, 182
   Transpersonal Human, 162, 173
air, land, and sea, 150, 156
alam al-mithal
   offspring D, 117
   Sufi, 117
Alexandra
   Revelations 135–136
ALICE
   non-locality, 13, 15–16, 18, 20, 23–24, 37, 123, 187
   non-ordinary Consciousness, 16–18
   twins, 16
ALL THAT IS
   God, 10–12, 14, 20, 35, 43–45, 50–51, 60, 64, 86, 87, 104–106, 109, 114, 116–118, 139, 155, 183, 191–192
   Source, 12, 15, 27, 38, 42, 64, 86–87, 99, 106, 108–109, 112–113, 115–116, 121–122, 125, 127, 134, 137, 148

ALL THAT IS—*Cont.*
   Spirit, 2, 19, 61–64, 67, 86–87, 91, 93, 96–99, 101–109, 111–113, 115, 118–120, 124–125, 128–129, 131, 133, 136–137, 139–141, 144, 147–148, 151, 157–159, 162, 164–166, 168–169, 174, 185, 193
alter
   Multiple Personality Disorder, 79, 94
Altered states of Awareness/Consciousness
   mystic, 3, 14, 16, 19, 27, 41, 79, 93, 99, 114, 117, 121–122, 125, 138–139, 187
   non-ordinary states, 23, 85–87, 91, 125
   Shaman, 4, 23, 62, 66, 117, 184, 191
Ananda, 92, 138, 144
anandamayokosha
   5 Kosha/Hindu offspring H, 118–120
ancient spiritual traditions, 3, 99
angry Separation of adolescence
anima
   archetypes, 22, 51–53, 59, 66, 88–89, 110, 122, 160, 163, 179, 191
   Jung, C.G., 22, 66, 191
   Patriarch, 54, 58, 73, 89, 91–92, 101, 125, 154–156, 159, 167, 175–177
animus
   archetypes, 22, 51–53, 59, 66, 88–89, 110, 122, 160, 163, 179, 191
   see also Jung, C.G.
Annie
   psychologist, 3, 78–79, 84–89, 91
anti-matter, 148
anti-particles, 148, 179
anxiety of separation, 71
APSYCHE, 110, 112, 114, 119, 125, 127, 131–132, 134, 138
archetype(s), *see also* anima, archetype (88)
   collective unconscious, 5–6, 51, 59, 63, 88, 91, 179, 181
   see also Jung, C.G.
"As above, so below"
   Jesus, 60, 114, 116, 135–136, 138
   macrocosm, 4, 20, 127
ASACRED MARRIAGE, 163–164
ascend and re-trace, 124–125
ASELF, 2, 70–94, 105, 110–112, 114, 126, 134–135, 138, 144–151, 153–166, 168, 173–174, 177, 182
ASOUL, 2–5, 108–110, 112–114, 118–120, 124, 126–128, 131–135, 138, 160, 164, 168, 172, 174
aspects and facets of personality
   inner children, 4, 51–52, 75, 88, 93, 145
   sub-personalities, 4, 51–52, 73, 75, 78–79, 88, 93
assumed authority
   mayan, 40, 113, 118
   offspring F, 113, 118
   Sorcerer, 118, 133, 191
ASUPER-IMPLICATE
   Bohm, D., 15, 16, 19–20, 37, 121–123
   Holomovement, 19–20, 37, 98, 106, 121–122, 124, 184
   World of Creation, 64, 109
at-one-ment, 88
attachment
   Childhood, 17, 28–29, 53, 92, 125, 147, 153–154, 160
   Communion, 29, 53, 58, 125, 147, 153–155, 177
Atziluth
   Jewish Mysticism, 67, 113–116
   offspring A, 114
   World of Origination, 109, 121–122, 124

AUNIVERSE, 127, 134
Australian Aborigines
   Dream Time, 11, 43, 134, 182
Australophithecus Afarenus
   ancestors, 2, 29–30, 128, 151
   H.Erectus, 151–152, 165
   H.Sapien, 151–152, 163, 165, 191
authority
   Mayan, 40, 113, 118
   offspring F, 118
autonomy, 54, 71, 76, 80, 90, 113, 118, 154, 167, 176, 179, 183
awakening spirit, 87, 98, 112
Aware-ized
   ALL THAT IS, 35, 43, 98, 102, 108–109, 111–112, 119, 121, 127–128, 144, 147, 161–163, 166, 179, 183, 189, 191
   Holomovement, 19–20, 37, 98, 106, 121–122, 124, 184
   intelligence, 75–76, 86, 105, 121–122, 127, 137, 148, 151
   Source, 12, 15, 27, 38, 42, 64, 86–87, 99, 106, 108–109, 112–113, 115–116, 121–122, 125, 127, 134, 137, 148
   SPIRIT, 2–3, 19, 61–64, 67, 86–87, 91, 93, 96–99, 101–105, 107–109, 111–115, 117–120, 124–125, 127–128, 131, 133, 136–137, 139–141, 144, 147–148, 151, 157–159, 162, 164–166, 168–169, 174, 185, 193
AWORLD
   Explicate, 15–16, 20, 105, 121–123, 183–184, 191
   World of Manifestation, 64, 98, 105, 111–116, 121–124, 130, 147, 150, 155

"Bad" Father/Patriarch, 50, 54, 58, 73, 89, 91–92, 101, 125, 154–156, 159, 167, 175–177
"Bad" Mother, 155 *see also* Section III
bardos, 119–120 133, 180,
BEAM ME DOWN, 135
before the before
   Big Bang, 10, 12–14, 20, 32, 180
   Creation, 3, 5–6, 10–14, 39, 41–42, 46, 50, 62, 64, 86, 92, 98–99, 105–111, 113–116, 118, 121–124, 126, 130–131, 133, 145, 147–148, 150–151, 157–158, 163–164, 177, 182
   Genesis, 11, 60, 107–108, 121, 147–148, 164
Bell's Theorem
   Non-Locality, 13, 15–16, 18, 20, 23–24, 37, 121–123, 187
   Quantum Mechanics, 4, 13, 15–16, 34, 35, 180–181, 185–186, 189, 192
belonged
   Communion, 29, 53, 58, 125, 147, 153–155, 177
   First People, 28–31, 33, 39–40, 117–118, 133, 147, 153–154
   Humankind, 2–3, 8–11, 25, 28–29, 31–40, 42–43, 50–51, 55, 57, 60, 82, 86, 88, 92, 101, 103, 106, 113, 116, 122, 125, 127–128, 135, 137, 146–157, 159–166, 168, 172–174, 176, 179, 181–182
Big Bang
   Creation, 3, 5–6, 10–14, 39, 41–42, 46, 50, 62, 64, 86, 92, 98–99, 105–111, 113–116, 118, 121–124, 126, 130–131, 133, 145, 147–148, 150–151, 157–158, 163–164, 177, 182
   genesis, 11, 60, 107–108, 121, 147–148, 164

195

birth and death
  cycles, 3, 12, 29–30, 40, 52–53, 58, 129, 150, 153, 167, 172, 193
  Dance of Shiva, 41, 164
  reincarnation, 37, 128–129, 189
birthright, 172
Blake, W., 136–137
blended family
  ASELF and Enough, 93, 144
  Blessing and Harmony, 84, 144
blueprints
  archetypes, 22, 51–53, 59, 66, 79, 88–89, 110, 122, 160, 163, 179, 191
  DNA, 18, 35, 88, 150, 181–183, 190
  Jung, C.G., 66 see also Jung
  M-fields, 13, 21–22, 121 see also Morphogenetic, fields, Sheldrake
  Sheldrake, R., 13, 21, 22, 121 - see also Sheldrake
  World of Formation, 64, 98, 110–111, 113–116, 126, 131, 134
Bohm, David
  Explicate, 15–16, 20, 105, 121, 123–124, 183–184, 191
  Holomovement, 19–20, 37, 98, 106, 121–122, 124, 184, 191
  Implicate, 13, 15–17, 20, 37, 45, 121–124, 126, 181, 183–185, 191
  Quantum Potential, 16, 20, 122, 189 - see also fields
  Super-Implicate, 121, 122
bonding
  Childhood, 17, 28–29, 53, 92, 147, 153–154, 160, 180
both/and mode of perception, 156, 166–168
  new paradigm, 9, 34, 37–39, 50, 59, 86, 121, 138, 167
bottom (beauty)
  quark, 4, 34–35, 44–46, 189
brain
  neurological-sensory apparatus, 13, 16, 25, 31, 52–53, 60–61, 81–82, 86–87, 99–100, 123, 125, 155
Brave New World, 2, 155, 168
BREATH OF THE INVISIBLE, 12, 157, see also Spirit
Buddhist
  Bardos, 120, 133, 180
  Clear Light, 37, 65, 113, 120, 180
  offspring I, 120
Butterfly Effect, 37–38

Cambridge
  Rising Star, 82–84
camouflage, 130
Capra, Fritjof
  Quantum Physics, 9, 36, 45, 139
Carpe Diem, 173
CATERPILLAR DREAMS
  Cagan, A., 161
  sacred chrysalis, 160–162
Catholic prelate
  offspring B, 114, 116
causal
  classical science, 2–4, 9, 28, 32–33, 81–82, 87, 180
  linear, 10, 26, 31–32, 38, 40, 125, 155, 157, 167, 180, 187
  old paradigm, 9–10, 33, 81, 167
cerebral hemispheres
  brain, 13, 17–18, 32, 37, 87, 95, 99, 169
  neurological apparatus, 13, 25, 125
chakra energy
  Kundalini energy, 117, 187
  offspring C, 117
chaos
  Chaos Theory, 20, 34, 42

chaos—Cont.
  complexity, 9, 20, 27, 34, 37, 42, 45–46, 52, 147, 150, 157, 159
  crisis, 29, 34, 38, 54–55, 73, 154, 156
  Quantum Mechanics, 4, 13, 15–16, 34, 181, 186, 189, 192
  uncertainty, 2, 4, 9, 28, 34, 36–38, 42–43, 84, 147, 156, 159, 167, 173, 177, 192
Chief Seattle
  interconnectivity, 20, 43
childhood
  ASELF, 2, 70–94, 105, 110–112, 114, 126, 134–135, 138, 144–151, 153–166, 168, 173–174, 177, 182
  Communion, 29, 53, 58, 117, 125, 147, 153–155, 177
  First People, 28–31, 33, 39–40, 117–118, 133, 147, 153–154
  Great Goddess, 53, 154, 175
  Matriarch, p.
choice
  Patriarch, 54, 58, 73, 89, 91–92, 101, 125, 154–156, 159, 167, 175–177
  Responsibility, 6, 55, 65, 90, 118, 127, 129, 133, 145–147, 154, 159–160, 165, 176, 190
  Separation, 28, 31, 43, 53–56, 58, 71, 76–78, 83, 86, 90, 101, 125, 153–155, 159, 167, 176–177, 190
Chop wood and carry water
  ordinary, 5, 27, 101, 103, 118, 125, 132–133, 136, 164
chronos time
  linear, 10, 26, 31–32, 38, 40, 125, 155, 157, 167, 187
  sequential time, 10, 26, 31–32, 38, 40, 125, 155, 157, 167, 187
circle of time
  Sacred time, 27–28, 167–168
CIRCULAR-SPIRAL MATRIX
  sacred circle, 2, 29, 31, 33, 38–40, 42, 101, 141, 147, 153, 155, 157–159
  Sacred time, 27–31, 101, 167–168
civilizations
  epochs, 26, 28–29
  eras, 28–29, 153, 155, 177
clarity of mind
  meditation, 16, 27, 41, 91, 119, 180–181, 186
Classical
  Newtonian-Cartesian, 87, 167
  old paradigm science, 9–10, 33, 81, 167
  Science, 2–4, 6–15, 17, 19–29, 31–35, 37–41, 43–47, 50, 61–62, 64, 66, 76, 81–82, 86, 92–93, 98, 105, 107, 121, 127, 129, 139–141, 145, 155–156, 164, 177, 182, 187, 192–193
CLEAR LIGHT
  Buddhism, 37, 43, 65, 102, 113, 120, 138, 180, 193
  offsprings H&I, 118–119
closed or open Universe
  Hawking & Hartle Theory, 14, 32
co-creator
  partnership, 110, 147, 162–163, 166–168
collective unconscious - see also Jung and archetypes
Communion
  Childhood, 17, 28–29, 53, 92, 125, 147, 153–154, 160, 177
  First People, 28–31, 33, 39–40, 117–118, 133, 147, 153–154
  Great Mother, 54–56, 70, 92, 95, 117, 125, 153–155, 159–160, 167
community of inner Selves
  personal unconscious, 4, 51–52, 59, 73, 75, 79–80, 88, 185, 188

community of inner Selves—Cont.
  sub-personalities, see also aspects and facets of personality
compassionate awakening
  see also archetypes
  partnership, 110, 147, 162–163, 166–168
  principles, 3, 13, 20, 52–57, 58, 65, 99, 122, 160, 163, 173, 177, 190
complexes - see also Jung, archetypes, Aspects and facets of personality
  inner selves
  shadow, 72–73, 75, 79, 88, 156, 159, 167, 176, 191
  sub-personality
Complexity
  chaos, 9, 20, 28, 34, 37–38, 40, 42–43, 45–46, 52, 65, 79–80, 84, 89, 101, 147, 156, 159, 163–164, 173, 177, 180, 187
  crisis, 29, 34, 38, 54–55, 73, 85, 154, 156
  Quantum Mechanics, 13, 27, 34, 181, 186, 189, 192
composite "I"
  Multiple Personality Disorder, 79, 94
composite "WE"
  Multiple Personality Disorder, 79, 94
concept of synchrony
  Jung, C.G., 22
  Sheldrake, R.
Consciousness, 2–6, 11, 13, 16, 19–21, 23–25, 27, 29–31, 36–37, 39, 42–43, 46, 51–58, 60, 62, 64–65, 67, 73, 79, 85–87, 91–92, 95, 98–100, 103, 105–109, 111–112, 117–120, 122–127, 129–130, 135, 137–138, 140, 144–148, 150, 153–157, 159–160, 162–167, 172–173, 175, 177, 179–183, 185–189, 191–193
continuous cycle of annihilation and creation
  Birth & Death, 189, 153–154, 161–162
  Dance of Shiva, 41, 164
  E=MC², 41, 164
  Evolution, 1, 3, 5–6, 9, 53, 55, 57, 62, 67, 88, 91, 93, 95, 98–99, 104–107, 112–114, 118, 122, 125, 127, 129, 135, 137, 139–140, 147–148, 150–152, 154, 159–160, 162, 165–168, 173, 177, 182, 185–186, 188, 190–191
continuum of gross, subtle and causal states
  Buddhist, offspring I, 120
  Hindu, offspring H, 118–120
  Koshas, 113, 118–120, 185
  Vedanta, 118–119
  Wilber, K., 65–66 see also Spectrum of Consciousness and Great Chain of Being
contrasexual archetypes
  anima, 89
  animus, 89
  collective unconscious, 5–6, 22, 51, 59, 63, 88, 91, 179, 181
  see also Jung and archetypes
core SELF, 64
cosmic map
  mystic, 3, 14, 16–17, 19, 27, 41, 79, 93, 99, 114, 117, 121, 125, 138–139, 187
  Path of the Heart, 93, 98–99, 103–104, 124, 127, 137, 144, 157, 160, 162, 165
cosmogenesis
  15 Billion years, 180, 148, 151–152
  Big Bang, 10, 12–14, 20, 32, 180
  genesis, 11, 60, 107–108, 121, 147–148, 164
  Hawkings & Hartle, 14, 32
Creation
  Cycles, 3, 12, 29–30, 40, 52–53, 58, 129, 150, 153, 167, 172, 193
  Dance of Shiva, 41, 164
creation stories
  Dream-Time, 11, 43, 66, 133–134

Index

creation stories—*Cont.*
 Four Worlds, 62, 98, 106, 108–116, 118, 121, 124–125, 137, 150, 177
 genesis, 11, 60, 107–108, 121, 147–148, 164
Creator
 ALL THAT IS, 35, 43, 98, 102, 108–109, 111–112, 119, 121, 127–128, 144, 147, 161–163, 166, 179, 183, 189, 191
 God, 10–12, 14, 20, 35, 43–45, 50–51, 60, 64, 66, 87, 104–106, 109, 114, 116–118, 139, 155, 183, 191–192
 Source, 12, 15, 27, 38, 42, 64, 86–87, 99, 106, 108–109, 112–113, 115–116, 121–122, 125, 127, 134, 137, 148
 SPIRIT, 2–3, 19, 61–64, 67, 86–87, 91, 93, 96–99, 101–105, 107–109, 111–115, 117–120, 124–125, 127–128, 131, 136–137, 139–141, 144, 147–148, 151, 157–159, 162, 164–166, 168–169, 174, 185, 193
crisis
 chaos & opportunity, 20, 34, 42
 Chinese, 14, 20, 29, 34, 38, 54–55, 73, 85, 154, 156, 192
critical developmental stage, 154, 54–56, 71
Cro-Magnon
 ancestors, 2, 29–30, 128, 151
 evolution, 1, 3, 5–6, 9, 53, 55, 57, 62, 67, 88, 91, 93, 95, 98–99, 104–107, 112–114, 118, 122, 125, 127, 129, 135, 137, 139–140, 147–148, 150–152, 154, 159–160, 162, 165–168, 173, 177, 182, 185–186, 188, 190–191
crucial developmental stages, 54–56, 71, 154
cyberspace
 illusion, 16, 18, 36, 130, 186
 Joe, 130–131
 "Maya", 16, 110, 130, 186
 virtual reality, 130–131, 135, 139, 181, 192
cycle of birth and death
 Dance of Shiva, 41, 162, 164
 $E=MC^2$, 41
 karma, 37, 128–129, 185
 reincarnation, 37, 128–129, 189
cycles
 First People, 28–31, 33, 39, 117–118, 133, 147, 153–154
 harmony, 28–31, 39, 57, 84, 92, 101, 144–145, 147, 158, 160, 163, 167–168, 172, 175
 Mother Earth, 3, 12, 28–30, 39–40, 52–53, 56, 70, 92, 149, 153, 155, 161, 167, 173
 rhythms, 12, 29, 31, 64, 78, 150, 153, 155, 172

DANCE OF SHIVA, 41, 164
dance score
 World of Formation, 64, 110–111, 124, 126, 131, 112, 98, 121–122
dancing Universe, 41, 161
dark side
 ego consciousness, 55–56, 64, 130, 154–155, 167 *see also* separation
 see also Jung, C.G.
 shadow, 72–73, 75, 79, 88, 156, 159, 167, 176, 191
Daughter
 ASELF, 2, 70–94, 105, 110–112, 114, 126, 134–135, 138, 144–151, 153–166, 168, 173–174, 177, 182
Davies, Paul, 13, 23, 32, 35, 43–45
Dean of Neuroscience
 ASELF, 2, 70–94, 105, 110–112, 114, 126, 134–135, 138, 144–151, 153–166, 168, 173–174, 177, 182

death of the advisor
 Mayan, 40, 113,118
 offspring F, 118
 Sorcerer, 118, 133, 191
defiant child
 disowned, 51–52, 55, 58, 72–73, 75, 79, 88, 145, 156, 185
 inner children, 51–52, 75, 88, 145
 split off, 51, 55, 73, 155, 159, 185
 sub-personalities, 51–52, 73, 75, 78–79, 88
Demeter - *see also* p. 175–176
 Goddess, 53–54, 56–58, 66–67, 70, 92, 147, 154–155, 160, 167, 169, 175–177
denser layers
 Four Worlds, 62, 98, 106, 108–116, 118, 121, 124–125, 137, 150, 177
 Great Chain of Being, 62, 64, 98, 106, 119, 125, 129, 137, 150, 159, 162, 177
 Matter, 2, 7–10, 13, 15–21, 29, 31–37, 39–44, 47, 50, 61, 64, 85–87, 99, 105–106, 109, 111, 113, 121–124, 128, 130, 133, 137, 148–149, 153, 157–158, 160, 162–164, 172, 179–183, 185–189, 191, 193
 spectrum of consciousness, 46, 58, 65, 85, 98, 106, 118–120, 125, 137–138, 150, 154, 159, 177, 179, 183, 185–186, 191
dependency needs vs. autonomy
 Communion and Separation, 28–29, 31, 53, 58, 71, 76–78, 83, 86, 90, 101, 125, 147, 153–155, 177, 190
descend
 evolution, 1, 3, 5–6, 9, 53, 55, 57, 62, 67, 88, 91, 93, 95, 98–99, 104–107, 112–114, 118, 122, 125, 127, 129, 135, 137, 139–140, 147–148, 150–152, 154, 159–160, 162, 165–168, 173, 177, 182, 185–186, 188, 190–191
 Four Worlds, 62, 98, 106, 108–116, 118, 121, 124–125, 137, 150, 177
destiny
 cross roads, 164
 World of 2070 A.D., 168, 172
deterministic
 Classical, 2–4, 6–15, 17, 19–29, 31–35, 37–41, 43–47, 50, 61–62, 64, 66, 76, 81–82, 86, 92–93, 98, 105, 107, 121, 127, 129, 139–141, 145, 155–156, 164, 152, 180, 184–185, 187–189, 191–3
 Ego Consciousness, 55–56, 130, 154–155, 167
 Patriarch, 54, 58, 73, 89, 91–92, 101, 125, 154–156, 159, 167, 175–177
 Separation, 28, 31, 43, 53–56, 58, 71, 76–78, 83, 86, 90, 101, 125, 153–155, 159, 167, 176–177, 190
devas
 elves, 50
 fairies, 50
developmental stages
 Childhood, 17, 28–29, 53, 92, 125, 147, 153–154, 160, 177
 evolution, 1, 3, 5–6, 9, 53, 55, 57, 62, 67, 88, 91, 93, 95, 98–99, 104–107, 112–114, 118, 122, 125, 127, 129, 135, 137, 139–140, 147–148, 150–152, 154, 159–160, 162, 165–168, 173, 177, 182, 185–186, 188, 190–191
 Separation, 28, 31, 43, 53–56, 58, 71, 76–78, 83, 86, 90, 101, 125, 153–155, 159, 167, 176–177, 190
 Spectrum of Consciousness, 46, 58, 65, 85, 98, 106, 118–120, 125, 137–138, 150, 154, 159, 177, 179, 183, 185–186, 191
dialogue
 inner selves, 4, 52, 72–73, 79, 92, 144, 185 *see also* Jung, archetypes, aspects and facets

dialogue—*Cont.*
 sub-personalities, 51–52, 73, 75, 78–79, 88
 Voices, 4, 59, 66, 73, 76–79, 81
deities
 ALL THAT IS, 10–12, 14, 20, 35, 43, 50–51, 60, 64, 66, 70, 87, 98, 101, 108–109, 111–112, 114, 116–118, 119, 121, 127–128, 139, 144, 147, 161–163, 166, 179, 183, 189, 191–192
 SOURCE, 12, 15, 27, 38, 42, 64, 86–87, 99, 106, 108–109, 112–113, 115–120, 124–125, 127, 134, 137, 139–141, 144, 148, 152, 158, 179, 183, 193
 SPIRIT, 2–3, 19, 61–64, 67, 86–87, 91, 93, 96–99, 101–105, 107–109, 111–115, 117–121, 123–125, 127–129, 131, 133, 135–137, 139–141, 144, 146–148, 151, 157–160, 162, 164–166, 168–169, 174, 179, 185, 189, 193
differentiation
 Ego-Consciousness, 55–56, 130, 154–155, 167
 Masculine, 53–58, 73, 89, 99, 101, 125, 147, 160, 162–163, 172–173, 177, 190, 193
 Patriarch, 54, 58, 73, 89, 91–92, 101, 125, 154–156, 159, 167, 175–177
 Separation, 28, 31, 43, 53–56, 58, 71, 76–78, 83, 86, 90, 101, 125, 153–155, 159, 167, 176–177, 190
dimensions of existence
 Four Worlds, 62, 98, 106, 108–116, 118, 121, 124–125, 137, 150, 177
 Great Chain of Being, 62, 64, 98, 108, 119, 125, 129, 137, 150, 159, 162, 177
 Realms of Spirit, 62, 96–99, 101, 103, 106, 108–109, 111, 112, 118, 122, 124–125, 127, 132, 133, 137, 158, 164–165, 172–173, 179, 182, 187, 191
 Spectrum of Consciousness, 46, 58, 65, 85, 98, 106, 118–120, 125, 137–138, 150, 154, 159, 177, 179, 183, 185–186, 191
direct and experiential
 gnosis, 87, 129, 136–137, 183
 Numinous, 82, 85, 92–93, 99, 101, 103, 106, 125, 131, 136–137, 160, 187
 receptive mode, 99–101
 revelation, 116, 135, 177, 190
 SPIRIT, 2–3, 19, 61–64, 67, 86–87, 91, 93, 96–99, 101–105, 107–109, 111–115, 117–121, 123–125, 127–129, 131, 133, 135–137, 139–141, 144, 146–148, 151, 157–160, 162, 164–166, 168–169, 174, 179, 185, 189, 193
disappearance
 evolution, 1, 3, 5–6, 9, 53, 55, 57, 62, 67, 88, 91, 93, 95, 98–99, 104–107, 112–114, 118, 122, 125, 127, 129, 135, 137, 139–140, 147–148, 150–152, 154, 159–160, 162, 164–168, 173, 177, 182, 185–186, 188, 190–191
 extinct, 39
 perish, 147, 166, 176
discrete order of time
 chronos time, 26
 linear, 10, 26, 31–32, 38, 40, 125, 155, 157, 167, 187
 separation, 28, 31, 43, 53–56, 58, 71, 76–78, 83, 86, 90, 101, 125, 153–155, 159, 167, 176–177, 190
Disowned and split off aspects
 inner children, 51–52, 75, 88, 145
 shadow, 72–73, 75, 79, 88, 156, 159, 167, 176, 191
 sub-personalities, 51–52, 73, 75, 78–79, 88

197

Dissipative structures, 9
  Chaos & Uncertainty, 20, 34, 37–38, 40, 42–43, 45–46, 52, 65, 84, 88–89, 101, 147, 149, 156, 163–164, 173, 177, 180, 187
  Prigogine, I., 9, 20, 28, 34
  Stengers, I., 28
diversity
  both/and, 156, 166–168
  circle-spiral, 30–31, 33, 38–40, 42, 101, 141, 147, 153, 155, 157, 167
  new paradigm science, 9, 34, 37–39, 50, 86, 121, 138, 167
Divine
  ALL THAT IS, 35, 43, 98, 102, 108–109, 111–112, 119, 121, 127–128, 144, 147, 161–163, 166, 179, 183, 189, 191
  God, 10–12, 14, 20, 35, 43–45, 50–51, 60, 64, 66, 87, 104–106, 109, 114, 116–118, 139, 155, 183, 191–192
  NUMINOUS, 82, 85, 92–93, 99, 101, 103, 106, 125, 131, 136–137, 160, 187
  SOURCE, 12, 15, 27, 35, 38, 42, 64, 86–87, 99, 106, 108–109, 112–113, 115–116, 119–122, 125, 127, 131, 137, 148, 152, 158, 179, 183, 193
  SPIRIT, 2–3, 19, 61–64, 67, 86–87, 91, 93, 96–99, 101–105, 107–109, 111–115, 117–120, 124–125, 127–128, 131, 133, 135–137, 139–141, 144, 146–148, 151, 157–160, 162, 164–166, 168–169, 174, 185, 193
divorce
  ALL THAT IS, 35, 43, 98, 102, 108–109, 111–112, 119, 121, 127–128, 144, 147, 161–163, 166, 179, 183, 189, 191
  ASELF, 2, 70–94, 105, 110–112, 114, 126, 134–135, 138, 144–151, 153–166, 168, 173–174, 177, 182
  Rising Star, 84–85, 89
DNA molecule
  blueprints, 18, 21, 35, 51–52, 80, 110–112, 122, 131, 133, 150, 181–183, 190
Doctrine of the Trinity
  Catholic prelate—offspring B, 114, 116
domain
  levels, 2, 5, 15–16, 25, 29, 34, 57–64, 70, 85–86, 88, 94, 98–99, 102, 104–109, 111–112, 115–116, 118–125, 128, 130, 134–135, 137, 150, 157, 159, 163, 177, 181, 183–185, 188–190, 192
  realms, 14, 19–20, 24–25, 27, 29, 57, 59, 61–63, 82, 85, 87, 91–92, 96–98, 101, 103, 106, 109–112, 118–119, 121, 123–127, 133, 137, 139, 141, 158, 165, 172–173
dominion
  Ego Consciousness, 55–56, 130, 154–155, 167
  Patriarch, 54, 58, 73, 89, 91–92, 101, 125, 154–156, 159, 167, 175–177
  Separation, 28, 31, 43, 53–56, 58, 71, 76–78, 83, 86, 90, 101, 125, 153–155, 159, 167, 176–177, 190
dream
  6 Bardos, 119–120, 133, 180,
  ASELF, 2, 70–94, 105, 110–112, 114, 126, 134–135, 138, 144–151, 153–166, 168, 173–174, 177, 182
  Australian Aborigines, 11, 133, 134, 182
  Carlos Castaneda, 133, 139
  Dream Self, 70, 72–73, 93, 134
  Dream-Time, 11, 43, 66, 133–134, 182
  Dream Universe, 132–133
  Dream Yogi, 133
  Dreams 1–19 see also Section six
  Dreamers, 172, 146

dream—Cont.
  lucid dreamers, LaBerge Stephen , 133
  Vishna 134
  Wolf, Fred A., 23–24, 44, 67, 121, 126, 133, 140
drive toward autonomy
  ASELF, 2, 70–94, 105, 110–112, 114, 126, 134–135, 138, 144–151, 153–166, 168, 173–174, 177, 182
  Ego Consciousness, 55–56, 130, 154–155, 167
  independence, 54, 71, 74, 76, 80, 90, 113, 118, 147, 153, 154, 155, 167, 176, 179, 183
  Masculine, 53–58, 73, 89, 99, 101, 125, 147, 160, 162–163, 172–173, 177, 190, 193
  Patriarch, 54, 58, 73, 89, 91–92, 101, 125, 154–156, 159, 167, 175–177
  psyche, 1–6, 9, 27, 42, 50, 52–59, 61–62, 65, 79, 87–91, 94, 98–99, 101, 106, 108, 110, 112–113, 115, 137–138, 145, 148, 150–151, 154–168, 172–173, 175–177, 181, 189–191, 147
duality, 3–4, 9, 56, 62, 86, 99, 137, 156, 167, 182, 193
Dukkha
  Buddhism, 37, 43, 65, 102, 113, 118–120, 133, 138, 180, 193
  suffering, 120, 182

$E=MC^2$
  annihilation & creation, 41, 164
  Dance of Shiva, 41, 164
  Einstein, A., 2–4, 6–15, 17, 19–29, 31–35, 37–41, 43–47, 50, 61–62, 64, 66, 76, 81–82, 86, 92–93, 98, 105, 107, 121, 127, 129, 139–141, 145 see also Special Theory of Relativity
  energy, 9, 16, 21, 26, 40–42, 52, 59, 109, 114, 117, 121–122, 130, 133, 148, 164, 172, 179–181, 183, 185, 187–189, 193
  mass, 15, 41, 179–180, 186–187
  matter, 2, 7–10, 14–15, 21, 29, 31–37, 39–44, 47, 50, 61, 64, 85–87, 99, 105–106, 109, 111, 113, 121–124, 128, 130, 133, 137, 148–149, 153, 157–158, 160, 162–164, 172, 179–182, 185–189, 191, 193
Earthling
  ASELF, 2, 70–94, 105, 110–112, 114, 126, 134–135, 138, 144–151, 153–166, 168, 173–174, 177, 182
  First People, 28–31, 33, 39, 40, 117–118, 133, 147, 153–154
Eastern traditions
  Pathways H and I, 118–120
ecology, 30, 39, 67, 140, 169, 172, 182
edge of chaos
  complexity, 9, 13, 20, 27, 28, 29, 34, 37, 38, 40, 42, 43, 45–46, 52, 65, 79–80, 84, 89, 101, 147, 150, 156, 157, 159, 163–164, 173, 177, 180, 186–187, 189, 192
  crisis, 14, 20, 29, 34, 38, 42, 54–55, 73, 85, 154, 156, 192
  Dance of Shiva, 41, 164
  $E=MC^2$, 41
  uncertainty, 2, 4, 9, 13, 19–20, 28, 34, 36–38, 40, 42–43, 45–46, 52, 65, 84, 89, 101, 147, 150, 156–157, 159, 163–164, 167, 173, 177, 180–181, 186–187, 189
editing
  ASELF, 2, 70–94, 105, 110–112, 114, 126, 134–135, 138, 144–151, 153–166, 168, 173–174, 177, 182

editing—Cont.
  ASOUL, 2–5, 108–110, 112–114, 118–120, 124, 126–128, 131–135, 138, 160, 164, 168, 172, 174
  life-scripts, 131, 133, 135
  Maya, 16, 18, 36, 110, 130, 131, 186
  virtual reality, 130–131, 135, 139, 181, 192
egg and sperm
  physical birth, 153, 162
Ego Consciousness
  independence, 54, 71, 74, 76, 80, 90, 113, 118, 147, 153, 154, 155, 167, 176, 179, 183
  individuation, 31, 54, 57, 77–78, 80–81, 88, 90, 95, 154, 185, 190
  Patriarch, 54, 58, 73, 89, 91–92, 101, 125, 154–156, 159, 167, 175–177
  Separation, 28, 31, 43, 53–56, 58, 71, 76–78, 83, 86, 90, 101, 125, 153–155, 159, 167, 176–177, 190
Ego Self
  independence, 54, 71, 74, 76, 80, 90, 113, 118, 147, 153, 154, 155, 167, 176, 179, 183
  individuation, 31, 54, 57, 77–78, 80–81, 88, 90, 95, 154, 185, 190
  Patriarch, 54, 58, 73, 89, 91–92, 101, 125, 154–156, 159, 167, 175–177
  Separation, 28, 31, 43, 53–56, 58, 71, 76–78, 81, 83, 86, 90, 101, 125, 153–155, 159, 167, 176–177, 190
ego-centric view
  transpersonal Self, 91–93, 156
ego/Patriarchal Consciousness
  autonomy, 54, 71, 76, 80, 90, 113, 118, 154, 167, 176, 179, 183
  see also Freud, Sigmund
  independence, 54, 71, 74, 76, 80, 90, 113, 118, 147, 153, 154, 155, 167, 176, 179, 183
  Oedipal Crisis/Electra Complex, 54–55, 154, Appendix III
  Separation, 28, 31, 43, 51, 53–56, 58, 71, 76–78, 81, 83, 86, 90, 94, 101, 125, 153–155, 159, 167, 176–177, 190
EGYPT
  non-locality, 13, 15–16, 18, 20, 23–24, 37, 123, 187
  non-ordinary experience, 16, 24
Eight levels of Consciousness
  Spectrum of Consciousness, 46, 58, 65, 85, 98, 106, 118–120, 125, 137–138, 150, 154, 159, 177, 179, 183, 185–186, 191
  Pathways H and I, 118–120
  Vedanta, Hindu, 119–120
  Wilber, K., 65–66
eight vijnanas, 120
eight-fold path of Buddha's Dharma
  Buddhism, 37, 43, 67, 102, 120, 138, 180, 193
  Wilber, K., 65–66
Ein-Sof
  Four Worlds, 62, 98, 106, 108–116, 118, 121, 124–125, 137, 150, 177
  God, 10–12, 14, 20, 35, 43–45, 50–51, 60, 64, 66, 87, 104–106, 109, 114, 116–118, 139, 155, 183, 191–192
  Jewish mysticism, 114
  Kabbalah, 113–116, 138–140, 185
  Pathway A, 114
Einstein's Special Theory of Relativity
  gravity, 12, 35, 182, 186, 189, 192
  non-local, 15, 18, 20, 22–23, 24, 37, 123, 187
  science, 2–4, 6–15, 17, 19–29, 31–35, 37–41, 43–47, 50, 61–62, 64, 66, 76,

Index

Einstein's Special Theory of Relativity—*Cont.*
science—*Cont.*, 81–82, 86, 92–93, 98,
105, 107, 121, 127, 129, 139–141, 145,
155–156, 164, 177, 182, 187, 192–193
space-time, 2, 4, 13, 18, 20, 22–23,
26–27, 35, 41, 47, 61, 70, 99, 107, 121,
123, 127–128, 130, 133, 153, 155, 162,
182–184, 186–187, 189, 191, 193
Electra Complex
see also Freud, Sigumund
Crisis of Oedipus, Freud, 54
Oedipal Complex, 54–55, 73, 154,
Appendix III
electron
non-local phenomenon, 13, 15–16, 18,
20, 23–24, 37, 123, 187
sub-atomic particles, 9, 34–36, 148–149,
179, 183, 186, 188–189, 191–192
eleventh century—offspring, 120
elves
devas, 50
fairies, 50
emergence of the Feminine
section five, 55, 62, 97, 159, 167
emergent properties
chaos, 9, 20, 28, 34, 37–38, 40, 42–43,
45–46, 52, 65, 79–80 84, 89, 101, 147,
156,159, 163–164, 173, 177, 180, 187
complexity, 9, 20, 27, 34, 37, 42, 45–46,
52, 79–80, 147, 150, 157, 159, *see also*
Complexity
dissipative structures, 20, 38, 181–182
Prigogine, I., 20, 28, 34
Stengers, I., 28
emerging Feminine
healer, 160
herald, 160
intuition, 56, 58, 163
receptive, 57, 73, 93, 99, 101, 160, 167,
179, 193
sacred marriage, 3, 57–59, 89, 101, 125,
138, 147, 163, 168, 172, 177, 190
transpersonal, 57–59, 62–66, 85, 87–88,
91–94, 98, 101, 112–113, 118, 125,
140, 145, 152, 160, 162–168, 173, 177,
190, 192, 194
empathetic
emerging feminine, 91, 93, 155, 160, 163
empowerment
emerging feminine, 91, 93, 155, 160, 163
sacred marriage, 3, 57–59, 89, 101, 125,
138, 147, 163, 168, 172, 177, 190
equivalence
Dance of Shiva, 41, 162, 164
E=MC^2, 41
emptiness
Buddhism, 37, 43, 67, 102, 120, 138, 180,
193
circle-spiral matrix, 2, 27, 29, 31, 33,
38–40, 101, 141, 147, 153, 155, 157–159
Jung, C.G., 66
energy
Dance of Shiva, 41, 162, 164
E=MC^2, 41
Mass, 15, 41, 179–180, 186–187
Matter, 2, 7–12, 15, 21, 29, 31–37, 39–44, 47,
50, 61, 64, 85–87, 99, 105–106, 109,
111, 113, 121–124, 128, 130, 133, 137,
148–149, 153, 157–158, 160, 162–164,
172, 179–182, 185–189, 191, 193
energy patterns
complexes, *see also* Jung, archetypes,
aspectes and facts
inner children, 51–52, 75, 88, 145, 185
sub-personalities, 51–52, 73, 75, 78–79, 88
enfolded
Bohm, D., *see also* Bohm, David
implicate & explicate, 191, *see also*
Bohm, David
holomovement, 19–20, 37, 98, 106,
121–122, 124, 184

enfolded—*Cont.*
levels, 2, 5, 15–16, 25, 29, 34, 57–64, 70,
85–86, 88, 94, 98–99, 102, 104–109,
111–112, 115–116, 118–125, 128, 130,
134–135, 137, 150, 157, 159, 163, 177,
181, 183–185, 188–190, 192
Spectrum of Consciousness, 46, 58, 65, 85,
98, 106, 118–120, 125, 137–138, 150,
154, 159, 177, 179, 183, 185–186, 191
enlightenment
awakening, 57, 66, 85, 87, 93, 98, 112,
114, 129, 133, 137, 149, 160–162, 164,
167, 177
mystic, 3, 14, 16–17, 19, 27, 41, 79, 93,
99, 114, 117, 121, 125, 138–139, 187
Path of the Heart, 93, 98–99, 103–104,
124, 127, 137, 144, 157, 160, 162, 165
Realms of Spirit, 62, 96–99, 101, 103,
106, 108–109, 111, 112, 118, 122,
124–125, 127, 132, 133, 137, 158,
164–165, 172–173, 179, 182, 187, 191
Self-realization, 59, 88, 91, 98, 104, 116,
125
ENOUGH
Ananda, 92, 144
ASELF, 2, 70–94, 105, 110–112, 114,
126, 134–135, 138, 144–151, 153–166,
168, 173–174, 177, 182
blended family, 93, 144
Canada, 92, 145
entelechy
actualization, 31, 109, 117, 126
archetypes, 22, 51–53, 59, 66, 88–89,
110, 122, 160, 163, 179, 191
blueprints, 13, 21, 51–52, 110–112, 122,
131, 133
Plato, 14, 51, 173–174, 183
see also Morphogenetic forms
entropy
see also Classical Science
Newton, 9, 87, 167, 179–180
Enuma Elish
Babylonia, 10
Genesis, 11, 60, 107–108, 121, 147–148,
164
environmental crises
Chaos & Uncertainty, 34–39
cross-roads, 164
warning signs, 156
equivalence
Dance of Shiva, 41, 162, 164
E=MC^2, 41
energy, 9, 11, 16, 21, 26, 40–42, 52, 59,
79, 109, 114, 117, 121–122, 130, 133,
148, 164, 172, 179–181, 183–189, 191,
193
Matter, 2, 7–12, 15, 21, 29, 31–37, 39–44,
47, 50, 61, 64, 85–87, 99, 105–106, 109,
111, 113, 121–124, 128, 130, 133, 137,
148–149, 153, 157–158, 160, 162–164,
172, 179–182, 185–189, 191, 193
Era
Communion, 29, 53, 58, 125, 147,
153–155, 177
epochs, 26, 28
Separation, 28, 31, 43, 53–56, 58, 71,
76–78, 83, 86, 90, 101, 125, 153–155,
159, 167, 176–177, 190
see also Chaos & Uncertainty, 34–39
see also Union & Reconnectivity, 40
erase personal history
existential, 4, 59, 62, 64, 90, 93, 98,
112–113, 118, 125, 145, 147, 156, 159,
162, 177
Mayan, 40, 113, 118
Nagual, 118, 187
offspring F, 118
tonal, 118, 192

Eros
archetypes, 22, 51–53, 59, 66, 88–89,
110, 122, 160, 163, 179, 191
Feminine, 53–58, 74–75, 89, 91–94, 99,
101, 117, 125, 147, 155, 159–160,
162–163, 167, 172–173, 175–177, 185,
190, 193
essential components of matter
physics, 9, 23, 43–46, 62, 76, 98, 106,
121, 139, 152, 189
Quantum Mechanics, 4, 13, 15–16, 34,
181, 186, 189, 192
sub-atomic particles, 9, 34–36, 148–149,
179, 183, 186, 188–189, 191–192
eternal
ALL THAT IS, 35, 43, 98, 102, 108–109,
111–112, 119, 121, 127–128, 144, 147,
161–163, 166, 179, 183, 189, 191
DIVINE, 51, 99, 102, 104–105, 107, 109,
116, 129, 136, 155, 161, 179
GOD, 10–12, 14, 20, 35, 43–45, 50, 60,
64, 66, 87, 104–106, 109, 114,
116–118, 139, 155, 183, 191–192
SELF, 2–6, 8–9, 17–18, 22–24, 27–28,
31, 45, 50, 52–56, 58–59, 63–64, 66,
70, 72–74, 79–80, 86–93, 98–99, 101,
104, 106, 108–110, 112, 117–118, 125,
127–132, 134–139, 141, 144–147,
154–155, 158, 162, 165, 168, 185,
188–191
SOUL, 43, 45, 55, 58–64, 66, 85, 98, 102,
105–111, 113–114, 117, 119, 127, 132,
137, 148, 160, 164, 180, 189–191
SOURCE, 12, 15, 27, 35, 38, 42, 64,
86–87, 99, 106, 108–109, 112–113,
115–116, 119–122, 125, 127, 131, 137,
148, 152, 158, 179, 183, 193
SPIRIT, 2–3, 19, 61–64, 67, 86–87, 91,
93, 96–99, 101–105, 107–109,
111–115, 117–120, 124–125, 127–128,
131, 133, 135–137, 139–141, 144,
146–148, 151, 157–160, 162, 164–166,
168–169, 174, 185, 193
eternal present
NOW, 10, 27–28, 37, 80, 89, 93, 107,
127, 136, 153, 161, 179, 192
simultaneous time, 101, 125, 128, 167, 191
timeless, 10–11, 14, 30, 40, 54, 109, 122,
125, 153, 166, 179, 182, 186
eternal validity of Consciousness
ALL THAT IS, 35, 43, 98, 102, 108–109,
111–112, 119, 121, 127–128, 144, 147,
161–163, 166, 179, 183, 189, 191
SOUL, 43, 45, 55, 58–64, 66, 85, 98, 102,
105–111, 113–114, 117, 119, 127, 132,
137, 148, 160, 164, 180, 189–191
Eukaryotes
Dreams - *see also* Section VI
evolution, 150–152
space, 160
Evolution of the Psyche
ASELF, 2, 70–94, 105, 110–112, 114,
126, 134–135, 138, 144–151, 153–166,
168, 173–174, 177, 182
Four Worlds, 62, 98, 106, 118, 121,
124–125, 137, 150, 177
Great Chain of Being, 62, 64, 98, 108,
119, 125, 129, 137, 150, 159, 162, 177
humankind, 2–3, 8–11, 25, 28–29, 31–40,
42–43, 50–51, 55, 57, 60, 86, 92, 101,
103, 106, 113, 116, 122, 125, 127–128,
133, 146–157, 159–166, 168, 172–176,
179, 181–182
levels, 2, 5, 15–16, 25, 29, 34, 57–64, 70,
85–86, 88, 94, 98–99, 102, 104–109,
111–112, 115–116, 118–125, 128, 130,
134–135, 137, 150, 157, 159, 163, 177,
181, 183–185, 188–190, 192

199

Evolution of the Psyche—*Cont.*
  Pathways, 3, 20–21, 23, 51, 91, 104, 113, 124–125, 157, 164
  Realms of Spirit, 62, 96–99, 101, 103, 106, 108–109, 111, 112, 118, 122, 124–125, 127, 132, 133, 137, 158, 164–165, 172–173, 179, 182, 187, 191
  Spectrum of Consciousness, 46, 58, 65, 85, 98, 106, 118–120, 125, 137–138, 150, 154, 159, 177, 179, 183, 185–186, 191
  Wilber, K., 65–66
evolutionary dead-end
  cross-roads, 164
  destiny, 42, 136, 157, 159, 166, 173, 179
  extinct, 39
  perish, 147, 166, 176
Ex-Centric Patriarch, 156, 163
  Evolution of Psyche, 1–3, 5, 53, 62, 88, 98–99, 101, 105–106, 113, 125, 135, 137, 118, 150, 154, 162, 165, 168
  process of transformation, 58, 125
  Wilber, K., 65–66
Existential
  ASELF, 2, 70–94, 105, 110–112, 114, 126, 134–135, 138, 144–151, 153–166, 168, 173–174, 177, 182
  awareness, 2, 4, 6, 20, 23, 27–28, 30, 39, 53–58, 60, 85, 87, 90, 92, 98–99, 101, 105–107, 109, 112–113, 117–122, 125, 127, 130–131, 133, 137, 145, 147, 155–157, 159, 162–163, 166, 168, 176, 180, 184–186, 190
  death & dying, 25
  Evolution of Psyche, 118, 150, 154, 162, 165, 168
  Spectrum of Consciousness, 46, 58, 65, 85, 98, 106, 118–120, 125, 137–138, 150, 154, 159, 177, 179, 183, 185–186, 191
  Wilber, K., 65–66
Existential Self
  ASELF, 2, 70–94, 105, 110–112, 114, 126, 134–135, 138, 144–151, 153–166, 168, 173–174, 177, 182
  death & dying, 25
  Evolution of Psyche, 118, 150, 154, 162, 165, 168
  Spectrum of Consciousness, 46, 58, 65, 85, 98, 106, 118–120, 125, 137–138, 150, 154, 159, 177, 179, 183, 185–186, 191
  Wilber, K., 65–66
expanded Personhood
  ASELF, 2, 70–94, 105, 110–112, 114, 126, 134–135, 138, 144–151, 153–166, 168, 173–174, 177, 182
  awakening, 57, 66, 85, 87, 93, 98, 112, 114, 129, 133, 137, 149, 160–162, 164, 167, 177
  Consciousness, 2–6, 11, 13, 16, 19–21, 23–25, 27, 29–31, 36–37, 39, 42–43, 46, 51–58, 60, 62, 64–65, 67, 73, 79, 85–87, 91–92, 95, 98–100, 103, 105–109, 111–112, 117–120, 122–127, 129–130, 135, 137–138, 140, 144–148, 150, 153–157, 159–160, 162–167, 172–173, 175, 177, 179–183, 185–189, 191–193
  Evolution of Psyche, 118, 150, 154, 162, 165, 168
  Spectrum of Consciousness, 46, 58, 65, 85, 98, 106, 118–120, 125, 137–138, 150, 154, 159, 177, 179, 183, 185–186, 191
EXPLICATE
  see also Bohm, D.
  Four Worlds, 62, 98, 106, 118, 121, 124–125, 137, 150, 177

EXPLICATE—*Cont.*
  Holomovement, 19–20, 37, 98, 106, 121–122, 124, 184
  implicate, 13, 15–17, 20, 37, 45, 121–124, 126, 181, 183–185, 191
  manifest world, 111, 118, 122, 132, 185
extinction
  altered states of consciousness, 27, 87
  cross-roads, 164
  destiny, 42, 136, 157, 166, 173, 179
  extraordinary
  non-ordinary, 16–17, 23–24, 85–87, 91, 125
  perish, 147, 166, 176
  Prokaryote, 149–150, 152
  species, 2, 6, 28–29, 33, 39, 42, 52, 65, 92, 127, 149–151, 154, 156, 166, 173, 184, 188
  warning signs, 156

facilitating Self
  ASELF, 2, 70–94, 105, 110–112, 114, 126, 134–135, 138, 144–151, 153–166, 168, 173–174, 177, 182
  inner child, 4, 52, 88, 185
  multi-facet, 2–3, 5, 8, 50–52, 59, 61, 70, 72–73, 76, 79, 88, 90–93, 105, 119, 156, 166–167, 190
  personal consciousness, 29, 88, 92, 94
  sub-personality, 79
fairies
  deva, 50
  elves, 50
false self
  ASELF, 2, 70–94, 105, 110–112, 114, 126, 134–135, 138, 144–151, 153–166, 168, 173–174, 177, 182
  inner children, 51–52, 75, 88, 145
  multi-faceted, 2–3, 5, 8, 18, 52, 59, 61, 70–72, 75, 79, 81, 88, 90, 92, 105, 119, 156, 166–167, 190
  shadow, 72–73, 75, 79, 88, 90, 156, 159, 167, 176, 191
  sub-personality, 79
Family of Selves
  inner selves, 4, 52, 72–73, 79, 92, 144
  multi-faceted, 2–3, 5, 8, 18, 52, 59, 61, 70, 72, 75, 79, 81, 88, 90, 92, 105, 119, 156, 166–167, 190
  personal unconsciousness, 51, 79, 94, 188, 190
  shadow, 72–73, 75, 79, 88, 156, 159, 167, 176, 191
family patterns
  ASELF, 2, 70–94, 105, 110–112, 114, 126, 134–135, 138, 144–151, 153–166, 168, 173–174, 177, 182
faster-than-the-speed-of-light
  Einstein, 9, 15, 21, 35, 186, 189
  non-locality, 13, 15–16, 18, 20, 23–24, 37, 121, 123, 187
Father
  ASELF, 2, 70–94, 105, 110–112, 114, 126, 134–135, 138, 144–151, 153–166, 168, 173–174, 177, 182
Father Sky
  childhood, 17, 28–29, 53, 92, 125, 147, 153–154, 160
  Communion, 29, 53, 58, 125, 147, 153–155, 177
  First People, 28–31, 33, 39–40, 117–118, 133, 147, 153–154
  Mother Earth, 12, 28–30, 39–40, 52–53, 56, 70, 92, 149, 153, 155, 161, 167, 173
feedback
  cycle of life, 52, 150, 173

feedback—*Cont.*
  nature, 2, 8–11, 19, 21, 28–40, 42–43, 50, 53, 60, 70, 79, 107, 150–151, 159, 173, 175–176
  system, 13, 15, 20, 25, 32, 38, 81, 87, 95, 114, 117, 119, 130, 152–153, 180–182, 184–185, 187–188, 190, 193
female and male deities
  Appendix I, 57, 175
  archetypes, 51–53, 59, 66, 88–89, 163, 179, 191
  collective unconsciousness, 5–6, 22, 51, 59, 63, 88, 94, 181
  Gods & Goddesses, 50
female Psyche
  animus, 89
  Electra Complex, 55
  Jung, C.G., 66, 88
feminine see also Emerging Feminine
  archetype, 53–54, 88, 179
  evolution of psyche, 118, 150, 154, 162, 165, 168
  intuitive, 50, 53, 58, 99, 101, 160, 167, 190
  transpersonal, 56–59, 62–66, 85, 87–88, 91–94, 98, 101, 112–113, 118, 125, 140, 145, 152, 160, 162–168, 173, 177, 190, 192, 194
  relational, 91, 93, 167
  receptive, 57, 73, 93, 99, 101, 160, 167, 179, 193
feminine anima
  contra-sexual archetype, 89
  Jung, C.G., 22, 66, 191
  male psyche, 89
  ASELF, 2, 70–94, 105, 110–112, 114, 126, 134–135, 138, 144–151, 153–166, 168, 173–174, 177, 182
Feminine principle
  archetype, 22, 51–53, 53–54, 59, 66, 88, 89, 110, 122, 160, 163, 179, 191
  healer, 160
  redeemer, 160
fermions
  physics, 9, 23, 43–46, 62, 76, 106, 121, 139, 152, 189, 98
  Quantum Mechanics, 4, 13, 15–16, 34, 181, 186, 189, 192,
  sub-atomic particles, 9, 34–36, 148–149, 179, 183, 186, 188–189, 191–192
fifth dimension
  hyperspace, 9, 14, 126, 152, 184
  Rod Serling, 166
  world of imagination, 168
First People
  childhood, 17, 28–29, 53, 92, 125, 147, 153–154, 160
  Communion, 29, 53, 58, 125, 147, 153–155, 177
  Earthling, 154–157, 159–161, 163–166, 168, 182
  harmony, 28–31, 39, 57, 84, 92, 101, 144–145, 147, 158, 160, 163, 167–168, 172, 175
  Kayapo´, 28, 30
  Mother Earth & Father Sky, 29, 30, 149, 153
five koshas
  Great Chain of Being, 62, 64, 98, 108, 119, 125, 129, 137, 150, 159, 162, 177
  Hindu, 16, 20, 113, 118–120, 134, 186
  offspring H, 118–119
  sheaths, 120
  Wilber, K., 65–66
FOREVER see also eternal present and external validity of consciousness
  enduring, 2–3, 11, 27, 30, 37, 86, 98, 123, 158, 185

Index

FOREVER—*Cont.*
  Eternal, 2–4, 10–11, 27, 30, 42, 64, 109, 112, 122, 125, 127, 130, 133, 153, 157–158, 165–166, 180, 191–192
form and pattern
  archetypes, 22, 51–53, 59, 66, 88–89, 110, 122, 160, 163, 179, 191
  DNA, 18, 35, 88, 150, 181–183, 190
  morphogenetic fields, 13, 21–22, 121, 122, 123, 135
  see also Jung, C.G.
  see also Sheldrake, R.
  world of formation, 64, 108, 110–111, 126, 131
formative cause
  morphogenetic fields, 13, 21–22, 121–122
  Sheldrake, R. *see also* M-fields, archetypes, DNA, least action pathways
  world of creation, 64, 109–110
four dimensional space-time
  physicality, 2, 15, 41, 90, 109, 111, 124, 126, 130–131, 135, 137, 148, 153, 157, 159, 163–164, 186
  world of manifestation, 64, 98, 105, 111–112, 113–116, 121–124, 130, 147, 150, 155,
Four Worlds
  see also Bohm, D., 15, 16, 19–20, 37, 121–123
  evolution, 1, 3, 5–6, 9, 53, 55, 57, 62, 67, 88, 91, 93, 95, 98–99, 104–107, 112–114, 118, 122, 125, 127, 129, 135, 137, 139–140, 147–148, 150–152, 154, 159–160, 162, 165–168, 173, 177, 182, 185–186, 188, 190–191
  Great Chain of Being, 62, 64, 98, 108, 119, 125, 129, 137, 150, 159, 162, 177
  Holomovement, 19–20, 37, 98, 106, 121–122, 124, 184
  Spectrum of Consciousness, 46, 58, 65, 85, 98, 106, 118–120, 125, 137–138, 150, 154, 159, 177, 179, 183, 185–186, 191
  Wilber, K., 65–66
Fourth Force of psychology
  depth psychology, 59
  Jungian Psychology, 67
  Transpersonal psyche, 98, 163
  Transpersonal, 57–59, 62–64, 85, 87–88, 91–93, 98, 101, 112–113, 118, 125, 152, 160, 162–168, 173, 177, 190, 192
  Wilber, K., 62, 65–66
  Washburn, M., 58
fragmentation, 86, 101, 144, 156
free choice
  Ego Consciousness, 55–56, 130, 154–155, 167 - *see also* Separation
  Patriarch/Masculine, 54, 58, 73, 89, 91–92, 99, 101, 125, 147, 154–156, 159, 160, 162–163, 167, 175–177, 172–173, 190, 193
  responsibility, 6, 55, 65, 90, 118, 127, 129, 133, 145–147, 154, 159–160, 165, 176, 190
Freud, Sigmund
  Appendix III
  Crisis of Oedipus, 54, 55, 154
  ego, 22, 55–56, 58–59, 64, 72–73, 86, 88, 90, 93, 98, 112–113, 118, 125, 128, 130, 134, 136, 145, 147, 154–155, 167, 177, 182, 184
  Electra Complex, 55
  id, 54, 72–73, 90, 93, 134, 184
  Psyche, 1–6, 9, 27, 42, 50, 52–59, 61–62, 65, 87–91, 94, 98–99, 101, 106, 108, 110, 112–113, 115, 137–138, 145, 148, 150–151, 154–168, 172–173, 175–177, 181, 186, 189–191, *see also* Evolution of the Psyche

Freud, Sigmund—*Cont.*
  Psychology, 2–4, 6, 13, 22, 39, 43–44, 47, 50, 58–62, 66–67, 69, 71, 73, 75, 77, 79, 81, 83, 85–87, 89, 91, 93–95, 98–99, 105, 107, 113, 118–120, 127, 129, 139–141, 164, 168, 177, 179, 181, 186, 187, 192–194
  super-ego, 54, 72–73, 93, 134, 191
fundamental Oneness
  ALL THAT IS, 35, 43, 98, 109, 111–112, 114, 119, 121, 127–128, 144, 147, 161–163, 166, 179, 189, 191
  field, 9, 15–16, 20, 22–23, 35, 42, 86, 106, 122, 124, 126, 183, 189, 191
  GOD, 10–12, 14, 20, 35, 43–45, 50, 51, 60, 64, 66, 87, 104–106, 109, 114, 116–117, 118, 139, 155, 183, 191–192
  ground, 3, 5, 12–13, 19, 23, 28, 61, 86, 94, 98, 103, 105–106, 111–112, 136–137, 157, 183–184, 193
  SPIRIT, 2, 19, 61–64, 67, 86–87, 91, 93, 96–99, 101–109, 111–113, 115, 118–120, 124–125, 128–129, 131, 133, 136–137, 139–141, 144, 147–148, 151, 157–159, 162, 164–166, 168–169, 174, 185, 193
  Transcendent, 23, 58–59, 64, 86, 98, 113, 168, 192
fundamental split
  evolution of psyche, 118, 150, 154, 162, 165, 168
  process of transformation, 58, 125
  psyche, 1–6, 9, 27, 42, 50, 52–59, 61–62, 65, 87–91, 94, 98–99, 101, 106, 108, 110, 112–113, 115, 137–138, 145, 150–151, 154–168, 172–173, 175–177, 181, 184, 189–191
  Spectrum of Consciousness, 46, 58, 65, 85, 98, 106, 118–120, 125, 137–138, 150, 154, 159, 177, 179, 183, 185–186, 191
future
  ASELF, 2, 70–94, 105, 110–112, 114, 126, 134–135, 138, 144–151, 153–166, 168, 173–174, 177, 182
  cross-roads, 164
  Cosmogenesis, 148, 152
  Destiny, 42, 136, 157, 159, 166, 173, 179
  evolution, 1, 3, 5–6, 9, 53, 55, 57, 62, 67, 88, 91, 93, 95, 98–99, 104–107, 112–114, 118, 122, 125, 127, 129, 135, 137, 139–140, 147–148, 150–152, 154, 159–160, 162, 165–168, 173, 177, 182, 185–186, 188, 190–191
  Galaxies, 148–149, 152
  inner children, 51–52, 75, 88, 145
  multi-faceted, 2–3, 5, 8, 18, 52, 59, 61, 70–72, 75, 79, 81, 88, 90, 92, 105, 119, 156, 166–167, 190
  personal unconsciousness, 51, 79, 88, 94, 190
  shadow, 72–73, 75, 79, 88, 156, 159, 167, 176, 191
  Transpersonal Human, 162, 173

Garden of Eden
  Great Mother, 54–56, 70, 92, 117, 125, 153–155, 159–160, 167
  Patriarch, 54, 58, 73, 89, 91–92, 101, 125, 154–156, 159, 167, 175–177
  Separation, 28, 31, 43, 53–56, 58, 71, 76–78, 83, 86, 90, 101, 125, 153–155, 159, 167, 176–177, 190
  Slay The Dragon, 54, 154
Garden of the Soul
  Pathway D, 117
  Sufi, 117

gateway
  ALL THAT IS, 35, 43, 98, 102, 108–109, 111–112, 119, 121, 127–128, 144, 147, 161–163, 166, 179, 183, 189, 191
  Feminine Principle, 56, 101, 160
  Numinous, 82, 85, 92–93, 99, 101, 103, 106, 125, 131, 136–137, 160, 187
  Path of the Heart, 59, 93, 98–99, 103–104, 124, 127, 137, 144, 157, 160, 162, 165
  Transpersonal, 57–59, 62–64, 85, 87–88, 91–93, 98, 101, 112–113, 118, 125, 152, 160, 162–168, 173, 177, 190, 192
gauge bosons
  Force, 35, 51, 59, 63–64, 94, 117, 179, 182–183, 185–186, 192
  Physics, 9, 23, 43–46, 62, 76, 106, 121, 139, 152, 189
  Quantum Mechanics, 13, 34, 181, 186, 189, 192
  sub-atomic particles, 9, 34–36, 148–149, 179, 183, 186, 188–189, 191–192
Genesis
  Big Bang, 10, 12–14, 20, 32, 180
  Cosmogenesis, 148, 152
  Creation, 3, 5–6, 10–14, 39, 41–42, 46, 50, 62, 64, 86, 92, 98–99, 105–111, 113–116, 118, 121–124, 126, 130–131, 133, 145, 147–148, 150–151, 157–158, 163–164, 177, 182
  Dream-Time, 11, 43, 66, 133, 134, 182
genetic blueprint
  DNA, 18, 35, 88, 150, 181–183, 190
Ghandi, M., 42, 64
gift of prophecy
  ASELF, 2, 70–94, 105, 110–112, 114, 126, 134–135, 138, 144–151, 153–166, 168, 173–174, 177, 182
  mystics, 3, 12–13, 16, 19–20, 27, 41, 79, 86, 98, 102, 107, 114, 117, 121, 125, 127, 130, 138–139, 164, 187–188
  seers, visions, *see also* mystics, section six
  World of 2070 A.D., 168, 172
global crises
  Chaos & Uncertainty, 34–39
  crises, 34, 38, 145–147
  cross-roads, 164
  warning signs, 156
gnosis, 87, 129, 136–137, 183
  direct, 52, 61, 94, 99, 102, 131, 133
  Realms of Spirit, 62, 96–99, 101, 103, 106, 108–109, 111, 112, 118, 122, 124–125, 127, 132, 133, 137, 158, 164–165, 172–173, 179, 182, 187, 191
  receptive mode, 99–101
  revelation, 116, 135, 177, 190
GOD
  ABSOLUTE, 14, 104, 106, 137, 182
  ALL THAT IS, 35, 43, 98, 109, 111–112, 114, 119, 121, 127–128, 144, 147, 161–163, 166, 179, 189, 191
  CREATOR, 11, 23, 29–30, 86, 109, 114, 123, 126–127, 137, 157
  SOURCE, 12, 15, 27, 38, 42, 64, 86–87, 99, 106, 108–109, 112–113, 115–116, 121–122, 125, 127, 134, 137, 148
  SPIRIT, 2, 19, 61–64, 86–87, 91, 93, 96–99, 101–109, 111–113, 115, 118–120, 124–125, 128–129, 131, 133, 136–137, 139–141, 144, 147–148, 151, 157–159, 162, 164–166, 168–169, 174, 185, 193
"God Particle" (Lederman, L), 35, 183
  Higgs Boson, 183
  Higgs Field, 35, 183
  theory of Everything, 35, 126
God, the Father
  Classical Science, 2–4, 9, 28, 32–32, 81–82, 87, 180

201

God, the Father—Cont.
  Ego Consciousness, 55–56, 64, 130, 154–155, 167 see also Separation
  Patriarch, 54, 58, 73, 89, 91–92, 101, 125, 154–156, 159, 167, 175–177
  Separation, 28, 31, 43, 53–56, 58, 71, 76–78, 83, 86, 90, 101, 125, 153–155, 159, 167, 176–177, 190
Goddess
  Childhood, 17, 28–29, 53, 92, 125, 147, 153–154, 160–177
  Communion, 28–29, 53, 58, 125, 147, 153–155, 177
  evolution of psyche, 118, 150, 154, 162, 165, 168
  First People, 28–31, 33, 39, 40, 117–118, 133, 147, 153–154
  Spectrum of Consciousness, 46, 58, 65, 85, 98, 106, 118–120, 125, 137–138, 150, 154, 159, 177, 179, 183, 185–186, 191
gods of Thunder and War, 52
Golden Age
  Communion, 29, 53, 58, 125, 147, 153–155, 177
  First People, 28–31, 33, 39, 40, 117–118, 133, 147, 153–154
  Garden of Eden, 54
  Goddess, 53–54, 56–58, 66–67, 70, 92, 147, 154–155, 160, 167, 169, 175–177
  Great Mother, 54–56, 70, 92, 117, 125, 153–155, 159–160, 167
  see also Matriarch
"Good" Father/Patriarch
  archetypes, 22, 51–53, 59, 66, 88–89, 110, 122, 160, 163, 179, 191
  ego Consciousness, 55–56, 64, 130, 154–155, 167
  Evolution of Psyche, 118, 150, 154, 162, 165, 168
  individuation, 31, 54, 57, 77–78, 80–81, 88, 90, 95, 154, 185, 190
  see also Separation
"Good" Mother
  archetypes, 22, 51–53, 59, 66, 88–89, 110, 122, 160, 163, 179, 191
  Communion, 29, 53, 58, 125, 147, 153–155, 177
  Evolution of Psyche, 118, 150, 154, 162, 165, 168
  Separation, 28, 31, 43, 53–56, 58, 71, 76–78, 83, 86, 90, 101, 125, 153–155, 159, 167, 176–177, 190
good-enough Mother, 70–71
Great Chain of Being
  Buddhism, 37, 43, 67, 102, 120, 138, 180, 193
  Hindu, 16, 20, 113, 118, 120, 134, 186
  Spectrum of Consciousness, 46, 58, 65, 85, 98, 106, 118–120, 125, 137–138, 150, 154, 159, 177, 179, 183, 185–186, 191
  Wilber, K., 62, 65–66
Great Fall
  Appendix III
  Crisis of Oedipus, 54–55
  Separation, 28, 31, 43, 53–56, 58, 71, 76–78, 83, 86, 90, 101, 125, 153–155, 159, 167, 176–177, 190
Great Mother/Goddess
  Childhood, 17, 28–29, 53, 92, 125, 147, 153–154, 160, 177
  Communion, 29, 53, 58, 125, 147, 153–155, 177
  First People, 28–31, 33, 39, 40, 117–118, 133, 147, 153–154
  see also Matriarch
  Slay the Dragon, 54, 154

guiding principles
  archetypes, 22, 51–53, 59, 66, 88–89, 110, 122, 160, 163, 179, 191
  organizing principles, 3, 52–53, 57, 62 - see also Evolution of the Psyche
hallowed covenant
  Communion, 29, 53, 58, 125, 147, 153–155, 177
  First People, 28–31, 33, 39, 40, 117–118, 133, 147, 153–154
  sanctuary, 153, 156
Hartle-Hawking Theory
  Closed or Open Universe, 14, 32
Hawking, Stephen
  Hawking-Hartle Theory, 14, 32
heal the split
  Psyche, 1–6, 9, 27, 42, 50, 52–59, 61–62, 65, 87–91, 94, 98–99, 101, 106, 108, 110, 112–113, 115, 137–138, 145, 150–151, 154–168, 172–173, 175–177, 181, 184, 189–191
  Spectrum of Consciousness, 46, 58, 65, 85, 98, 106, 118–120, 125, 137–138, 150, 154, 159, 177, 179, 183, 185–186, 191
healthy ego, 88
Heidegger, M., 28
Heisenberg's Principle of Uncertainty
  Quantum Mechanics, 13, 15, 16, 34, 36, 181, 186, 189, 192
Hero and Heroine's Journey
  Path of the Heart, 93, 98–99, 103–104, 124, 127, 137, 144, 157, 160, 162, 165
hidden dimensions
  Four Worlds, 62, 98, 106, 108, 116, 118, 121, 124–125, 137, 150, 177
  Great Chain of Being, 62, 64, 98, 108, 119, 125, 129, 137, 150, 159, 162, 177
  Mystic, 3, 14, 16–17, 19, 27, 41, 79, 93, 99, 114, 117, 121, 125, 138–139, 187
Higg's particle or field
  Lederman, L., 35, 183
  Theory of Everything, 35, 126, 183
higher levels of Awareness, 57, 125, 163
Hindu
  Koshas, 113, 118–120, 185
  offspring H, 118–119
holding environment
  Childhood, 17, 28–29, 53, 92, 125, 147, 153–154, 160
  Communion, 29, 53, 58, 125, 147, 153–155, 177
  Mothering Ones, 70–71, 172
hologram
  see also Bohm, D., 15, 16, 19–20, 37, 121–123
  Pribram, C., 23
HOLOMOVEMENT
  Bohm, D., 15, 16, 19–20, 37, 121–123
  Four World of Scientists, 120–124
  Quantum Potential, 16, 20, 122, 189 - see also fields
Holy Qur'an
  offspring D, 117
Holy Spirit
  offspring B, 114, 116
Homo Erectus
  ancestors, 2, 29–30, 128, 151–152, 165, 184
  evolution, 1, 3, 5–6, 9, 53, 55, 57, 62, 67, 88, 91, 93, 95, 98–99, 104–107, 112–114, 118, 122, 125, 127, 129, 135, 137, 139–140, 147–148, 150–152, 154, 159–160, 162, 165–168, 173, 177, 182, 185–186, 188, 190–191
Homo Habilis
  ancestors, 2, 29–30, 128, 151–152, 165, 184

Homo Habilis—Cont.
  evolution, 1, 3, 5–6, 9, 53, 55, 57, 62, 67, 88, 91, 93, 95, 98–99, 104–107, 112–114, 118, 122, 125, 127, 129, 135, 137, 139–140, 147–148, 150–152, 154, 159–160, 162, 165–168, 173, 177, 182, 185–186, 188, 190–191
Homo Sapien
  ancestors, 2, 29–30, 128, 151–152, 165, 184
  evolution, 1, 3, 5–6, 9, 53, 55, 57, 62, 67, 88, 91, 93, 95, 98–99, 104–107, 112–114, 118, 122, 125, 127, 129, 135, 137, 139–140, 147–148, 150–152, 154, 159–160, 162, 163, 165–168, 173, 177, 182, 185–186, 188, 190–191
Hopi Indians
  Creation stories, 11
  Four Worlds, 62, 98, 106, 108, 116, 118, 121, 124–125, 137, 150, 177
Human Consciousness
  human psyche, 2–3, 5, 9, 52–53, 56, 65, 94, 101
  Human species, 54, 58, 184
Humanity, 2–3, 6, 9–10, 22, 25, 28–29, 31–33, 37–38, 40, 42, 50, 65, 82, 88, 98, 106–107, 114, 127, 135, 141, 146–151, 153–157, 159–164, 166, 173, 192
Humanity's future
  section six & seven,
Humankind
  Communion, 29, 53, 58, 125, 147, 153–155, 177
  Covenant with Mother Earth/Father Sky, 153, 155, 161, 172
  First People, 28–31, 33, 39, 40, 117–118, 133, 147, 153–154
Humankind's ancestors
  Cro Magnon, 151–152, 165
  Homo Erectus, 151–152, 184
  Table 10, 165
Houston, Jean
  Goddesses, 50–52
  Gods, 2, 10, 31, 33, 50–52, 106, 118, 175–176
  Myths, 10, 30, 46, 50–52, 57, 101, 103, 107, 118
Hyperspace, 9, 14, 126, 152, 184

I AM
  ALL THAT IS, 35, 43, 98, 109, 111–112, 114–115, 119, 121, 127–128, 144, 147, 161–163, 166, 179, 189, 191
  SOURCE, 12–15, 27, 38, 42, 64, 86–87, 99, 106, 108–109, 112–113, 115–116, 121–122, 125, 127, 134, 137, 148
  SPIRIT, 2, 11, 19, 61–64, 67, 86–87, 91, 93, 96–99, 101–109, 111–115, 118–120, 124–125, 128–129, 131, 133, 136–137, 139–141, 144, 146–148, 151, 157–159, 162, 164–166, 168–169, 174, 185, 193
"I" vs."Thou"
  differentiation, 53–54, 160, 191
  Ego-Consciousness, 55–56, 64, 73, 130, 154–155, 167
  individuation, 31, 54, 57, 77–78, 80–81, 88, 90, 95, 154, 185, 190
  Patriarch, 54, 58, 73, 89, 91–92, 101, 125, 154–156, 159, 167, 175–177
  Separation, 28, 31, 43, 53–56, 58, 71, 76–78, 83, 86, 90, 101, 125, 153–155, 159, 167, 176–177, 190
id, 54, 72–73, 90, 93, 134, 184
  see also Freud, Sigmund

Index

ideas
  Super-implicate, 121
  World of Creation, 64, 109
identity
  Ego, 22, 55–56, 58–59, 64, 72–73, 86, 88, 90, 93, 98, 112–113, 118, 125, 128, 130, 134, 136, 145, 147, 154–155, 167, 177, 182, 184
  Persona, 72–73, 88, 188
  personality, 5, 50–51, 59, 70, 72–73, 79, 88, 93, 145, 181, 185–186, 188–191
  Separation, 28, 31, 43, 53–56, 58, 71, 76–78, 83, 86, 90, 101, 125, 153–155, 159, 167, 176–177, 190
illusion
  Joe, 130–131
  Maya, 16, 18, 36, 110, 130–131, 186
  Virtual Reality, 130–131, 135, 139, 181, 192
illusory world
  Joe, 130–131
  Maya, 16, 18, 36, 110, 130, 131, 186
  Virtual Reality, 130–131, 135, 139, 181, 192
Imaginal realms
  offspring E, 117
  Shaman, 4, 23, 62, 117, 184, 191
  Wolf, 23, 44, 67, 121, 126, 133, 140
Immersive Technology
  illusion, 16, 18, 36, 130, 186
  Maya, 16, 18, 36, 110, 130, 131, 186
  virtual reality, 130–131, 135, 139, 181, 192
impermanence of matter
  Buddhism, 37, 43, 65, 102, 113, 120, 138, 180, 193
  Clear Light, 37, 65, 113, 120, 180
  Illusion, 16, 18, 36, 130, 186
  Maya, 16, 110, 130, 186
IMPLICATE
  see also Bohm, D.
  Holomovement, 19–20, 37, 98, 106, 121–122, 124, 184
  see also Worlds of Creation and Formation, 64, 98, 110–111, 126, 131
incredible certainty
  gnosis, 87, 129, 136–137, 183
  mystical, 4, 15–16, 61–63, 66–67, 74, 91, 98, 102, 108–109, 113, 121, 136–137, 140, 146–147, 168, 185, 187
  non-ordinary, 16–17, 23–24, 85–87, 91, 125
  revelation, 116, 135, 177, 190
independent Personhood
  Ego-Consciousness, 55–56, 64, 73, 130, 154–155, 167
  Patriarch, 54, 58, 73, 89, 91–92, 101, 125, 154–156, 159, 167, 175–177
  Separation, 28, 31, 43, 53–56, 58, 71, 76–78, 81, 83, 86, 90, 101, 125, 153–155, 159, 167, 176–177, 190
India
  Koshas, 113, 118–120, 185
  offspring H, 118–119
Indian Sorcerer
  Art of Dreaming, 132–133, 139, 168, 191
  Maya, 16, 18, 36, 110, 130, 131, 186
  offspring F, 118
Indigenous
  Childhood, 17, 28–29, 53, 92, 125, 147, 153–154, 160, 177
  Communion, 29, 53, 58, 125, 147, 153–155, 167
  Covenant, 153, 155, 161, 172
  First People, 28–31, 33, 39, 40, 117–118, 133, 147, 153–154
  tribes, 92

individual's history
  Persona, 72–73, 88, 188
  Personality, 5, 50–51, 59, 70, 72–73, 79, 88, 93, 145, 181, 185–186, 188–191
  (*see also* sub-personalitites)
individuation
  Ego-Consciousness, 55–56, 73, 130, 154–155, 167
  Patriarch, 54, 58, 73, 89, 91–92, 101, 125, 154–156, 159, 167, 175–177
  Separation, 28, 31, 43, 51, 53–56, 58, 71, 76–78, 81, 83, 86, 90, 94, 101, 125, 153–155, 159, 167, 176–177, 190
ineffable
  ALL THAT IS, 35, 43, 98, 109, 111–112, 114, 119, 121, 127–128, 144, 147, 161–163, 166, 179, 189, 191
  Numinous, 82, 85, 92–93, 99, 101, 103, 106, 125, 131, 136–137, 160, 187
  SOURCE, 12, 15, 27, 38, 42, 64, 86–87, 99, 106, 108–109, 112–113, 115–116, 121–122, 125, 127, 134, 137, 148
  SPIRIT, 2, 19, 61–64, 67, 86–87, 91, 93, 96–99, 101–109, 111–113, 115, 118–120, 124–125, 128–129, 131, 133, 136–137, 139–141, 144, 147–148, 151, 157–159, 162, 164–166, 168–169, 174, 185, 193
infancy
  Childhood, 17, 28–29, 53, 92, 147, 153–154, 160, 180
  Communion, 29, 53, 58, 125, 147, 153–155, 177
  First People, 28–31, 33, 39, 40, 117–118, 133, 147, 153–154
  Great Mother, 54–56, 70, 92, 117, 125, 153–155, 159–160, 167
  see also Matriarch
  pre-personal, 54, 58, 74, 92, 154, 168
infinite Realms of SPIRIT
  Four Worlds, 62, 98, 106, 108, 116, 118, 121, 124–125, 137, 150, 177
  Great Chain of Being, 62, 64, 98, 106, 119, 125, 129, 137, 150, 159, 162, 177
  levels, 2, 5, 15–16, 25, 29, 34, 57–64, 70, 85–86, 88, 94, 98–99, 102, 104–109, 111–112, 115–116, 118–125, 128, 130, 134–135, 137, 150, 157, 159, 163, 177, 181, 183–185, 188–190, 192 (*see also* fields, ground)
  Spectrum of Consciousness, 46, 58, 65, 85, 98, 106, 118–120, 125, 137–138, 150, 154, 159, 177, 179, 183, 185–186, 191
inherited
  archetypes, 22, 51–53, 59, 66, 88–89, 110, 122, 160, 163, 179, 191
  Collective unconscious, 5–6, 22, 51, 59, 63, 88, 91, 179, 181
  DNA, 18, 35, 88, 150, 181–183, 190
  Jung, C.G., 66, 87, 89
inner cast-off, 55
Inner Child - *see also* aspects and facrts of personality
  archetype, 53–54, 88, 153, 179
  Jung, C.G., 66, 88
  shadow, 72–73, 75, 79, 88, 156, 159, 167, 176, 191
  sub-personalities, 51–52, 73, 75, 78–79, 88
inner paths
  Path of the Heart, 93, 98–99, 103–104, 124, 127, 137, 144, 157, 160, 162, 165
  Peace, 57, 60, 72, 89, 92, 136, 153
  Realm, 11, 13–15, 18, 21, 23, 31, 47, 51, 61–62, 64, 70, 73, 106, 109, 111–112, 118, 122, 124–125, 127, 132–133, 172, 179, 182, 187, 191 - *see also* Realms of Spirit

inner paths—*Cont.*
  Realms, 14, 19–20, 24–25, 27, 29, 57, 59, 61–63, 82, 85, 87, 91–92, 96–99, 101, 103, 106, 109, 111–112, 118, 124–127, 133, 137, 158, 164–165, 172–173
  selves, 3–5, 25, 33, 50, 52, 59, 72–73, 75, 79, 81, 90, 92–93, 109, 127, 135, 145
  transcendence, 23, 58, 64, 87, 98, 107, 112–113, 140, 160, 168, 172
insight and pain
  evolution, 1, 3, 5–6, 9, 53, 55, 57, 62, 67, 88, 91, 93, 95, 98–99, 104–107, 112–114, 118, 122, 125, 127, 129, 135, 137, 139–140, 147–148, 150–152, 154, 159–160, 162, 165–168, 173, 177, 182, 185–186, 188, 190–191
  transformation, 23, 50, 57–58, 60, 85, 88, 92, 103, 106, 125, 128–129, 156, 162, 186, 192
instantaneously
  non-locality, 13, 15–16, 18, 20, 23–24, 37, 123, 187
instinctual (id) desire, 54, 72–73, 90, 93, 134, 184
  see also Freud, Sigmund
intellect and intuition
  see also Ego-Consciousness
  Patriarch, 54, 58, 73, 89, 91–92, 101, 125, 154–156, 159, 167, 175–177
  rational, 31–33, 38, 47, 50, 74–75, 99, 101, 103, 155–156, 167
  Separation, 28, 31, 43, 53–56, 58, 71, 76–78, 83, 86, 90, 101, 125, 153–155, 159, 167, 176–177, 190
interface
  Aborigines, 11, 133
  Dream-Time, 11, 43, 66, 133, 134, 182
  psychology, 2–4, 6, 13, 22, 39, 43–44, 47, 50, 58–62, 66–67, 69, 71, 73, 75, 77, 79, 81, 83, 85–87, 89, 91, 93–95, 98–99, 105, 107, 113, 118–120, 127, 129, 139–141, 164, 168, 177, 179, 181, 186–187, 192–194
  sacred time, 27–28, 167–168
internal dualities
  anima & animus, 89 - *see also* archetypes
  archetypes, 22, 51–53, 59, 66, 88–89, 110, 122, 160, 163, 179, 191
  see also Masculine & Feminine
  sacred marriage, 3, 57–59, 89, 101, 125, 138, 147, 163, 168, 172, 177, 190
interpenetrates
  ALL THAT IS, 10–12, 14, 20, 35, 43, 50–51, 60, 64, 66, 87, 98, 102, 108–109, 111–112, 116–119, 121, 127–128, 139, 144, 147, 161–163, 166, 179, 183, 189, 191–192
  densities, 185
  levels, 2, 5, 15–16, 25, 29, 34, 57–64, 70, 85–86, 88, 94, 98–99, 102, 104–109, 111–112, 115–116, 118–125, 128, 130, 134–135, 137, 150, 157, 159, 163, 177, 181, 183–185, 188–190, 192
  SPIRIT, 2–3, 19, 61–64, 67, 86–87, 91, 96–99, 101–105, 107–109, 111–115, 117–121, 123–125, 127–129, 131, 133, 135–137, 139–141, 144, 146–148, 151, 157–160, 162, 164–166, 168–169, 174, 179, 185, 189, 193
  SOURCE, 12, 15, 27, 38, 42, 64, 86–87, 99, 106, 108–109, 112–113, 115–120, 124, 125, 127, 134, 137, 139–144, 148, 152, 158, 179, 183, 193
interpersonal skills
  see also feminine, 56, 101, 160
  transpersonal, 57–59, 62–64, 85, 87–88, 91–93, 98, 101, 112–113, 118, 125, 152, 160, 162–168, 173, 177, 190, 192

203

intimacy
    honesty, 85, 159
    transpersonal Self, 91–93
intuition
    Feminine, 53–58, 66, 74–75, 89, 91–94,
        99, 101, 117, 125, 147, 155, 159–160,
        162–163, 167, 172–173, 175–177, 185,
        190, 193
    holistic, 9, 118
    modes of perception, 99–100
    subjective, 12, 44, 47, 82, 118, 155
invisible dimension
    Four Worlds, 62, 98, 106, 108, 118, 121,
        124–125, 137, 150, 177
    Great Chain of Being, 62, 64, 98, 108,
        119, 125, 129, 137, 150, 159, 162, 177
    inner worlds, 72, 102, 117, 161
    levels, 2, 5, 15–16, 25, 29, 34, 57–64, 70,
        85–86, 88, 94, 98–99, 102, 104–109,
        111–112, 115–116, 118–125, 128, 130,
        134–135, 137, 150, 157, 159, 163, 177,
        181, 183–185, 188–190, 192
    realms, 14, 19–20, 24–25, 27, 29, 57, 59,
        61–63, 82, 85, 87, 91–92, 96–99, 101,
        103, 106, 109, 111–112, 118, 124–127,
        133, 137, 158, 164–165, 172–173
    Spectrum of Consciousness 46, 58, 65, 85,
        98, 106, 118–120, 125, 137–138, 150,
        154, 159, 177, 179, 183, 185–186, 191
Iron John, The Wild Man
    archetype, 53–54, 88, 153, 179
    Bly, R., 52, 60
isolation
    Separation, 28, 31, 43, 53–56, 58, 71,
        76–78, 83, 86, 90, 101, 125, 153–155,
        159, 167, 176–177, 190
Jacob's Ladder
    Jewish Mysticism, 114–116
    Kabbalah, 113–116, 185
    offspring A, 114–116
Jesus Christ the son
    As Above, So Below, 4, 60, 92
    Catholicism, 114, 116
    offspring B, 114, 116
    Revelation, 116, 135, 177, 190
Joe
    Maya, 16, 18, 36, 110, 130–131, 186
    Virtual Reality, 130–131, 135, 139, 181,
        192
journey
    ascent and evolution, 118, 125, 137, 160
    evolution, 1, 3, 5–6, 9, 53, 55, 57, 62, 67,
        88, 91, 93, 95, 98–99, 104–107,
        112–114, 118, 122, 125, 127, 129, 135,
        137, 139–140, 147–148, 150–152, 154,
        159–160, 162, 165–168, 173, 177, 182,
        185–186, 188, 190–191
    Great Chain of Being, 62, 64, 98, 108,
        119, 125, 129, 137, 150, 159, 162, 177
    Path of the Heart, 93, 98–99, 103–104,
        124, 127, 137, 144, 157, 160, 162, 165
    return, 4, 11, 17, 22, 35, 37, 40, 42–43,
        50, 55–56, 80, 83–84, 112, 115, 124,
        136, 160, 176, 179
    Spectrum of Consciousness, 46, 58, 65, 85,
        98, 106, 118–120, 125, 137–138, 150,
        154, 159, 177, 179, 183, 185–186, 191
Jung, Carl Gustav
    analysis, 64, 99, 104, 193
    anima & animus, 89
    archetypes, 22, 51–53, 59, 66, 88–89,
        110, 122, 160, 163, 179, 191
    blueprints, 18, 21, 35, 51–52, 88, 110–112,
        122, 131, 133, 150, 181–183, 190
    collective unconscious, 5–6, 22, 51, 59,
        63, 88, 91, 179, 181
    individuation & separation, 31, 54, 57,
        77–78, 80–81, 88, 90, 95, 154, 185, 190

Jung, Carl Gustav—*Cont.*
    Neuman, E., 53–55
    Oedipal Complex, 54–55, 73 - *see also*
        Slaying the Dragon, 154

Karma
    birth & death, 153, 154, 162–162, 189
    cycle, 37, 40, 52, 54, 94, 123–124,
        128–130, 149–151, 162, 164, 173, 189
    reincarnation, 37, 128–129, 189
Kayapo´ universe
    Communion, 29, 53, 58, 125, 147,
        153–155, 177
    First People, 28–31, 33, 39, 40, 117–118,
        133, 147, 153–154
knowing
    gnosis, 87, 129, 136–137, 183
    revelation, 116, 135, 177, 190
Kundalini
    Chakra, 113, 117, 180, 185, 187
    offspring C, 117

La Berge, S
    Art of Dreaming, 132–133, 139, 168, 191
    Bardos, 37, 120, 133, 180
    Castaneda, 66, 133, 132
    Wolf, 23, 44, 67, 121, 126, 133, 140
larger whole
    Bohm, D., 16, 19–20, 37, 98, 106,
        121–122, 126, 184, 189, 191
    Four Worlds, 62, 98, 106, 118, 121,
        124–125, 137, 150, 177
    Holomovement, 19–20, 37, 98, 106,
        121–122, 124, 184
Law of Karma, 129
laws of nature
    Classical, 9, 28, 32, 152, 180, 184–185,
        187–189, 191
    Newton, I., 9–10, 28, 31, 33, 43, 51, 53–56,
        58, 71, 76–78, 81, 83, 86, 90, 94, 101,
        125, 153–155, 159, 167, 176–177, 190
    Old Paradigm Science, 9–10, 32, 33, 81,
        167
Lawrence, D.H., 27
least action pathway
    see also Morphogenetic fields
    see also Sheldrake, R.
    see also Wolf, F.A.,
Lederman, L
    Quantum Mechanics, 13, 34, 35, 126,
        181, 183, 186, 189, 192
left hemisphere
    brain, 13, 17–18, 31–32, 37, 87, 99
    logical, 31, 50, 75, 155, 167
    neurological apparatus, 13, 25, 125
    rational, 31–33, 38, 47, 50, 74–75, 99,
        101, 103, 155–156, 167
Letting Go
    Healing The Split, 59, 87, 176 *see also*
        Evolution of the Psyche
    Revelation, 116, 135, 177, 190
level
    awareness, 2, 4, 6, 20, 23, 27–28, 30, 39,
        53–58, 60, 85, 87, 90, 92, 98–99, 101,
        105–107, 109, 112–113, 117–122, 125,
        127, 130–131, 133, 137, 145, 147,
        155–157, 159, 162–163, 166, 168, 176,
        180, 184–186, 190
    see also Bohm, D.
    Four Worlds, 62, 98, 106, 118, 121,
        124–125, 137, 150, 177
    Great Chain of Being, 62, 64, 98, 106,
        119, 125, 129, 137, 150, 159, 162, 177
    ground, 3, 5, 12–13, 19, 23, 28, 61, 86,
        94, 98, 103, 105–106, 111–112, 122,
        136–137, 157, 183–184, 193

level—*Cont.*
    Holomovement, 19–20, 37, 98, 106,
        121–122, 124, 184
    implicate, 13, 15–17, 20, 37, 45,
        121–124, 126, 181, 183–185, 191
    Spectrum of Consciousness, 46, 58, 65, 85,
        98, 106, 118–120, 125, 137–138, 150,
        154, 159, 177, 179, 183, 185–186, 191
    understanding, 2–3, 5–6, 8–9, 13, 16, 22,
        24, 27–32, 34, 39, 42–44, 50, 52, 60–65,
        74, 78, 80, 87, 99, 106, 112, 117, 121,
        127, 129, 131, 135, 137, 141, 147, 155,
        162–163, 165, 168, 180, 187, 190, 192
    Wilber, K., 62, 65–66
Life and Birth
    birth & death, 153, 154, 161–162, 189
    creation & annihilation, 42
    cycle, 37, 40, 52, 54, 94, 123–124,
        128–130, 149–151, 162, 164, 173, 189
    Dance of Shiva, 41, 162, 164
    karma, 37, 128–129, 185
    reincarnation, 37, 128–129, 189
life scripts, 131, 133, 135
    birth and death, 123, 128–129, 164, 189
    creation & annihilation, 42
    Dance of Shiva, 41, 162, 164
    karma, 37, 128–129, 185
    reincarnation, 37, 128–129, 189
lifetimes
    birth and death, 123, 128–129, 164, 189
    creation & annihilation, 42
    Dance of Shiva, 41, 162, 164
    karma, 37, 128–129, 185
    reincarnation, 37, 128–129, 189
linear arrow of time
    chronos time, 26
    see also Ego-Consciousness
    entropy, 32, 37, 182, 190
    Old Paradigm Science, 9–10, 32–33, 81,
        167
    Patriarch, 54, 58, 73, 89, 91–92, 101, 125,
        154–156, 159, 167, 175–177
    Separation, 28, 31, 43, 53–56, 58, 71,
        76–78, 83, 86, 90, 101, 125, 153–155,
        159, 167, 176–177, 190
logical
    Ego-Consciousness, 73
    Masculine, 53–58, 73, 89, 99, 101, 125,
        147, 160, 162–163, 172–173, 177, 190,
        193
    mode, 31, 74, 99, 101, 155–156
    Patriarch, 54, 58, 73, 89, 91–92, 101, 125,
        154–156, 159, 167, 175–177
    Separation, 28, 31, 43, 51, 53–56, 58, 71,
        76–78, 81, 83, 86, 90, 94, 101, 125,
        153–155, 159, 167, 176–177, 190
Lover and Beloved
    offspring D, 117
    Sufi, 117
lucid
    Art of Dreaming, 132–133, 168, 191
    Bardos, 119, 120, 133, 180
    Castaneda, C., 66, 132–133
    dreamer, 113, 118, 133–134, 154
    dreaming, 11, 110, 119–120, 126,
        132–134, 146, 168, 180, 191
    La Berge, S., 133
    Wolf, F.A., 23, 44, 67, 121, 126, 133, 140

macrocosm
    as above, so below, 4, 60, 92
    Jesus, 60, 114, 116, 135–136, 138
    microcosm, 4, 20, 127
magical thinking
    Childhood, 17, 28–29, 53, 92, 125, 147,
        153–154, 160
    superstitious, 82

Index

MAI
- meditation, 16, 27, 41, 91, 120, 180–181, 186
- yoga teacher, 91

mammals
- evolution, 1, 3, 5–6, 9, 53, 55, 57, 62, 67, 88, 91, 93, 95, 98–99, 104–107, 112–114, 118, 122, 125, 127, 129, 135, 137, 139–140, 147–148, 150–152, 154, 159–160, 162, 165–168, 173, 177, 182, 185–186, 188, 190–191

mandalas
- Center, 9, 20, 109, 117, 158, 190
- circle, 2, 5, 20, 26, 29–31, 33, 38–40, 42, 93, 101, 106, 120, 141, 143, 147, 153, 155, 157–159
- evolution, 1, 3, 5–6, 9, 53, 55, 57, 62, 67, 88, 91, 93, 95, 98–99, 104–107, 112–114, 118, 122, 125, 127, 129, 135, 137, 139–140, 147–148, 150–152, 154, 159–160, 162, 165–168, 173, 177, 182, 185–186, 188, 190–191
- Jung, C.G., 66, 157–158
- universal, 9, 54, 70, 86, 183, 187

manifest
- Explicate, 15–16, 20, 105, 121, 123, 124, 183–184, 191
- matter, 2, 7–12, 15–21, 29, 31–37, 39–44, 47, 50, 61, 64, 85–87, 99, 105–106, 109, 111, 113, 121–124, 128, 130, 133, 137, 148–149, 153, 157–158, 160, 162–164, 172, 179–182, 185–189, 191, 193
- physicality, 2, 15, 41, 90, 109, 111, 124, 126, 130–131, 135, 137, 148, 153, 157, 159, 163–164, 186
- Self, 2–6, 8–9, 17–18, 22–24, 27–28, 31, 45, 50, 52–56, 58–59, 63–64, 66, 70, 72–74, 79–80, 86–93, 98–99, 101, 104, 106, 108–110, 112, 117–118, 125, 127–132, 134–139, 141, 144–147, 154–155, 158, 162, 165, 168, 185, 188–191
- World of Manifestation, 64, 105, 98, 111–112, 113–116, 121–124, 130, 147, 150, 155,

Manifestation
- physical world, 10, 13, 15–16, 18–19, 21–23, 31, 51, 56, 60–61, 79–81, 92, 99, 104, 111, 123–124, 128, 130, 132–133, 137, 192

manomayakosha
- Buddhism, 37, 102, 120, 180, 193
- Kosha, 113, 118–120, 185
- Hindu, offspring H, 118–120

map for personal
- guide, xxi, 3, 22, 24–25, 51–53, 59, 74, 84, 101, 103, 106, 139, 148, 168
- mystics, 3, 12–13, 16, 19–20, 27, 41, 79, 86, 98, 102, 107, 114, 117, 121, 125, 127, 130, 138, 164, 187–188

Marlboro Man
- archetype, 52–54, 88, 153, 179

Marriage
- ASELF, 1–2, 70–94, 105, 110–112, 114, 126, 134–135, 138, 144–151, 153–166, 168, 173–174, 177, 182
- sacred marriage, 3, 57–59, 89, 101, 125, 138, 147, 163, 168, 172, 177, 190

Mary
- Alexandra, 135–136
- Revelation, 116, 135, 177, 190
- Virgin Mother, 135–136

masculine *see also* Patriarch
- archetypal principle, 54–56, 175
- Patriarch, 8, 9, 54, 58, 73, 89, 91–92, 99, 101, 125, 147, 154–156, 159, 160, 162–163, 167, 172–173, 175–177, 190, 193

Masculine and Feminine
- archetypes, 22, 51–53, 59, 66, 88–89, 110, 122, 160, 163, 179, 191
- Ego-Consciousness, 55–56, 130, 154–155, 167
- Jung, C.G., 66
- Patriarch, 54, 58, 73, 89, 91–92, 101, 125, 154–156, 159, 167, 175–177
- Separation, 28, 31, 43, 53–56, 58, 71, 76–78, 83, 86, 90, 101, 125, 153–155, 159, 167, 176–177, 190

masculine animus
- feminine psyche, 88, 89, 91–93, 159–160, 162–164

mass-energy
- density, 111–112, 128
- energy, 9, 11, 16, 21, 26, 40–42, 52, 59, 79, 109, 114, 117, 121–122, 130, 133, 148, 164, 172, 179–181, 183–185, 189, 191, 193
- Manifestation, 21, 29, 37, 62, 64, 98, 105, 108, 111–114, 115, 116, 118, 121–124, 130, 147, 150, 155, 177,
- matter, 2, 7–12, 15–21, 29, 31–37, 39–44, 47, 50, 61–64, 85–87, 99, 105–106, 109, 111, 113, 121–124, 128, 130, 133, 137, 148–149, 153, 157–158, 160, 162–164, 172, 179–182, 185–189, 191, 193
- World of Physicality, 2, 41, 90, 111, 126, 137, 159, 164 *see also* World of Manifestation

Matriarch
- archetype, 53–54, 88, 153, 179
- Childhood, 17, 28–29, 53, 92, 125, 147, 153–154, 160, 177
- Communion, 29, 53, 58, 125, 147, 153–155, 177
- First People, 28–31, 33, 39, 40, 117–118, 133, 147, 153–154
- Goddess, 53–54, 56–58, 66–67, 70, 92, 147, 154–155, 160, 167, 169, 175–177
- Great Mother, 54–56, 70, 92, 117, 125, 153–155, 159–160, 167
- Pre-Personal, 54, 58, 74, 92, 154, 168

Matriarchy
- archetype, 53–54, 88, 153, 179
- Childhood, 17, 28–29, 53, 92, 125, 147, 153–154, 160, 177
- Communion, 29, 53, 58, 125, 147, 153–155, 177
- First People, 28–31, 33, 39, 40, 117–118, 133, 147, 153–154
- Goddess, 53–54, 56–58, 66–67, 70, 92, 147, 154–155, 160, 167, 169, 175–177
- Great Mother, 54–56, 70, 92, 117, 125, 153–155, 159–160, 167
- Pre-Personal, 54, 58, 74, 92, 154, 168

Matter
- physicality, 2, 15, 41, 90, 109, 111, 124, 126, 130–131, 135, 137, 148, 153, 157, 159, 163–164, 186
- World of Manifestation, 64, 105, 111–112, 115, 116, 124, 130, 147, 150, 155

Maya
- Buddhism, 37, 43, 65, 102, 113, 120, 138, 180, 193
- illusion, 16, 18, 36, 130, 186
- Joe, 130–131
- virtual reality, 130–131, 135, 139, 181, 192

Mayan
- offspring F, 118
- sorcerer, 118, 133, 191

mechanistic
- Classical, 9, 28, 32, 152, 180, 184–185, 187–189, 191
- Old Paradigm Science, 9–10, 32, 33, 81, 167

medicine wheels
- center, 9, 20, 109, 117, 158, 190
- circle-spiral matrix, 5, 27, 29, 31, 33, 38–40, 64, 101, 124, 141, 147, 153, 155, 157–159
- *see also* mandala, 117, 157–158
- Self, 2–6, 8–9, 17–18, 22–24, 27–28, 31, 45, 50, 52–56, 58–59, 63–64, 66, 70, 72–74, 79–80, 86–93, 98–99, 101, 104, 106, 108–110, 112, 117–118, 125, 127–132, 134–139, 141, 144–147, 154–155, 158, 162, 165, 168, 185, 188–191

meditation
- Mai, 91

Meister Eckehart
- mysticism, 27, 114, 187

microcosm
- As Above, So Below, 4, 60, 92
- Jesus, 60, 114, 116, 135–136, 138
- macrocosm, 4, 20, 127

middle world
- offspring E, 117
- Shaman, 4, 23, 62, 66, 117, 184, 191

mind
- offspring H & I, 118–120
- psyche, 1, 6, 9, 27, 42, 50, 52–59, 61–62, 65, 87–91, 94, 98–99, 101, 106, 108, 110, 112–113, 115, 137–138, 145, 148, 150–151, 154–168, 172–173, 175–177, 181, 189–191 *see also* Evolution of the Psyche

modern Homo Sapiens - 151–152, 163, 165
- ancestors, 2, 29–30, 128, 151–152, 165, 184
- evolution, 1, 3, 5–6, 9, 53, 55, 57, 62, 67, 88, 91, 93, 95, 98–99, 104–107, 112–114, 118, 122, 125, 127, 129, 135, 137, 139–140, 147–148, 150–152, 154, 159–160, 162, 165–168, 173, 177, 182, 185–186, 188, 190–191
- humankind, 2–3, 8–11, 25, 28–29, 31–40, 42–43, 50–51, 55, 57, 60, 86, 92, 101, 103, 106, 113, 116, 122, 125, 127–128, 133, 146–157, 159–166, 168, 172–176, 179, 181–182

Morphogenetic
- least action pathways, 21, 23, 185
- patterns, fields, forms, 13, 21–22, 121–122, *see also* patterns, fields, forms
- Sheldrake, R., 13, 21–22, 121–122, 186

most dense
- explicate, 15–16, 20, 105, 121, 123, 124, 183–184, 191
- levels, 2, 5, 15–16, 25, 29, 34, 57–64, 70, 85–86, 88, 94, 98–99, 102, 104–109, 111–112, 115–116, 118–125, 128, 130, 134–135, 137, 150, 157, 159, 163, 177, 181, 183–185, 188–190, 192 *see also* field, ground
- Spectrum of Consciousness, 46, 58, 65, 85, 98, 106, 118–120, 125, 137–138, 150, 154, 159, 177, 179, 183, 185–186, 191
- Wilber, K., 62, 65–66

most rarefied
- Four Worlds, 62, 98, 106, 108–116, 118, 121, 124–125, 137, 150, 177
- Great Chain of Being, 62, 64, 98, 106, 119, 125, 129, 137, 150, 159, 162, 177
- SPIRIT, 2–3, 19, 61–64, 86–87, 91, 93, 96–99, 101–105, 107–109, 111–115, 117–120, 124–125, 127–128, 131, 133, 136–137, 139–141, 144, 147–148, 151, 157–159, 162, 164–166, 168–169, 174, 185, 193

Mother
- ASELF, 2, 70–94, 105, 110–112, 114, 126, 134–135, 138, 144–151, 153–166, 168, 173–174, 177, 182

205

Once Upon ASOUL

Mother—*Cont.*
    caretaker, 53, 70, 153, 172
    Childhood, 17, 28–29, 53, 92, 125, 147, 153–154, 160, 177
    Communion, 29, 53, 58, 125, 147, 153–155, 177
    pre-personal, 54, 58, 74, 92, 154, 168
Mothering One, 53, 70–71, 153, 172
    implicate order
      see also Bohm, D.
    see also Holomovement, 15–16, 19–20, 37, 98, 106, 121–122, 124, 184
multi-faceted, 70–72
    inner selves, 4, 52, 72–73, 79, 92, 144
    see also sub-personalities, aspects and facets of personality
    nature, 2, 8–11, 19, 21, 28–40, 50, 53, 60, 70, 79, 107, 150–151, 159, 173, 175–176
    persona, 72–73, 88, 188
    personal unconscious, 4, 51–52, 59, 73, 75, 79–80, 88, 185, 188
    personality, 5, 50–51, 59, 70, 72–73, 79, 88, 93, 145, 181, 185–186, 188–191
    Personhood, 2–3, 5, 18, 28, 50, 52, 54, 59, 65, 72, 75, 79, 85, 90, 109–111, 134, 136, 141, 154, 158, 162 - *see also* Separation and Ego Consciousness
Multiple Personality Disorder (MPD)
    alters, 16, 79, 94
    Dissociative Identity Disorder, 79, 94
multiple universes
    parallel worlds, 125, 126
    Quantum World, *see also* Bohm, David Wolf, F.A., 23, 67, 126
    worlds, 2, 23, 61–62, 64, 72, 74, 92–93, 98–99, 108–113, 115–116, 118, 121–124, 130, 133–134, 137, 147, 163, 172, 191
mysteries
    ALL THAT IS, 10–12, 14, 20, 35, 43, 50–51, 60, 64, 66, 87, 98, 109, 111–112, 114, 116–119, 121, 127–128, 135–136. 138–139, 144, 147, 161–163, 166, 179, 189, 191–192
    Catholicism, 114, 116
    Christ, 55, 60, 114, 116
    Hidden, 13, 16, 27, 45, 51, 61, 63–64, 73, 79, 86, 101, 103, 106–107, 110, 115–116, 124, 127, 133, 137, 139, 146, 183–185, 190
    Jesus, 60, 114, 116, 135–136, 138
    offspring B, 114, 116
    SPIRIT, 2, 19, 61–64, 67, 86–87, 91, 93, 96–99, 101–109, 111–113, 115, 118–120, 124–125, 128–129, 131, 133, 136–137, 139–141, 144, 147–148, 151, 157–159, 162, 164–166, 168–169, 174, 185, 193
    SOURCE, 12, 15, 27, 38, 42, 64, 86–87, 99, 106, 108–109, 112–113, 115–116, 121–122, 125, 127, 134, 137, 148
    unknown, 30, 35, 91, 98, 118, 187
mystic
    eternal, 2–4, 10–11, 27, 30, 42, 64, 109, 112, 122, 125, 127, 130, 133, 153, 157–158, 165–166, 180, 191–192
    gnosis, 87, 129, 136–137, 183
    revelation, 116, 135, 177, 190
    Seer, 86–87, 117, 190
    Visionary, 117, 146,168
mystical
    ASELF, 2, 70–94, 105, 110–112, 114, 126, 134–135, 138, 144–151, 153–166, 168, 173–174, 177, 182
    Inner child, 4, 51–73, 52, 75, 78–79, 65, 88, 145, 185 *see also* facets of personality

mystical—*Cont.*
    Inner Selves, 4, 52, 72–73, 79, 92, 144
    void, 10–11, 13–14, 109, 121, 192
Myth of Oedipus, 54, 55, 73, 154, 187, Appendix III
    Crisis of Oedipus, 54, 55, 73, 154, Apendix III
    Freud, S., *see also* Appendix III
    Jung, C.G., 54, 66
    Neuman, E., 53–55
    Separation, 28, 31, 43, 51, 53–56, 58, 71, 76–78, 81, 83, 86, 90, 94, 101, 125, 153–155, 159, 167, 176–177, 190
    Steiner, R., 54–55

nadis
    Chakra, 113, 117, 180, 187
    Kundalini, 117, 185, 187
    offspring C, 117
NAGUAL
    Mayan, 40, 113, 118
    mysterious, 4, 85, 87, 101, 109, 117–118, 163, 187
    offspring F, 118
    unknown, 30, 35, 91, 98, 118, 187
Nature
    Mother Earth, 3, 12. , 28–30, 39–40, 52–53, 56, 70, 92, 149, 153, 155, 161, 167, 173
    nurture, 148, 172, *see also* Mothering Ones
    reality, 2–4, 10–12, 16, 18, 23, 25, 28–29, 56, 58, 62, 64, 72, 81–82, 86, 90, 92, 101, 106, 126–127, 130–131, 137, 155–156, 163, 167–168, 173–174
    rhythms, 12, 29, 31, 64, 78, 153, 155, 172
Neanderthal
    ancestors, 2, 29–30, 128, 151–152, 165, 184
    dream world, 73, 132–135
    Homo Sapien, 151–153, 162–163, 165
Nefesh
    Jewish Mysticism, 114–116
    offspring A, 114–116
    Soul level, 114
nervous system
    brain, 13, 17–18, 32, 37, 87, 95, 99, 169
    neurological, 13, 16, 25, 81, 125
Neshama
    Jewish Mysticism, 114–116
    offspring A, 114–116
    Soul level, 114
Neumann, E.
    Jung, C.G., 54, 66
    Masculine & Feminine Principles, 53–54
    Oedipus Crisis, 54–55, 73, 154, Appendix III
neurological apparatus
    brain, 13, 17–18, 32, 37, 61, 87, 95, 99, 169
    Right/Left Hemisphere, 32, 61
Neuroscience
    ASELF, 2, 70–94, 105, 110–112, 114, 126, 134–135, 138, 144–151, 153–166, 168, 173–174, 177, 182
    Dean, 91, 145
new levels of complexity
    Chaos, 9, 20, 28, 34, 37–38, 40, 42–43, 45–46, 52, 65, 84, 89, 101, 147, 156, 163–164, 173, 177, 180, 187
    dissipative structures, 20, 38, 181–182
    Prigogine, I., 20, 28, 34
new paradigm
    Both/And diversity, 3, 6, 37, 56, 151, 157–159, 167–168, 172
    New Science, 34, 44, 46, 93, 140

new Personhood
    evolution, 1, 3, 5–6, 9, 53, 55, 57, 62, 67, 88, 91, 93, 95, 98–99, 104–107, 112–114, 118, 122, 125, 127, 129, 135, 137, 139–140, 147–148, 150–152, 154, 159–160, 162, 165–168, 173, 177, 182, 185–186, 188, 190–191
    Transpersonal Human, 162, 173
newly emergent
    Chaos, 9, 20, 28, 34, 37–38, 40, 42–43, 45–46, 52, 65, 84, 89, 101, 147, 156, 163–164, 173, 177, 180, 187
    complexity, 9, 20, 27, 34, 37, 42, 45–46, 52, 79–80, 147, 150, 157, 159
    Prigogine, I, 20, 28, 34, 38, 181–182
    properties, 9, 20, 32, 36, 38, 40, 42, 62, 147, 150, 180, 182, 189, 193
Newtonian
    Classical, Deterministic, 9–10, 33, 81, 87, 167
    Old Paradigm Science, 2–4, 6–15, 17, 19–29, 31–35, 37–41, 43–47, 50, 61–62, 64, 66, 76, 81, 82, 86, 92–93, 167
    Separation, 28, 31, 43, 53–56, 58, 71, 76–78, 83, 86, 90, 101, 125, 153–155, 159, 167, 176–177, 190
No-When
    Big Bang, 10, 12–14, 20, 32, 148, 180
    space-time, 2, 4, 13, 18, 20, 22–23, 26–27, 35, 41, 47, 61, 70, 99, 107, 121–123, 127–128, 133, 153, 155, 162, 182, 184, 186–187, 189, 191
No-Where
    Big Bang, 10, 12–14, 20, 32, 148, 180
    space-time, 2, 4, 13, 18, 20, 22–23, 26–27, 35, 41, 47, 61, 70, 99, 107, 121–123, 127–128, 133, 153, 155, 162, 182–184, 186–187, 189
non-linear
    new paradigm, 9, 34, 37–39, 50, 59, 86, 121, 138, 167
    time, 2, 4–5, 8, 10–15, 18–22, 24–32, 40–42, 51, 61, 70, 77, 83, 84, 90–93, 101, 103, 107, 109–111, 117, 122–123, 125–128, 132–134, 144–146, 148–152, 155, 159, 161–162, 165, 167–168, 173, 179–180, 182–187, 189, 192, 193
    self-organizing properties, 9, 20, 32, 36, 38, 40, 42, 62, 147, 150, 180, 182, 189, 193
non-locality
    Alice, 16–18
    Bell's Theorem, 15
    Bohm, D., 37
    Egypt, 24–25
    electrons, 15, 123, 179, *see also* subatomic particles
non-material realm
    hidden realms, 27, 127
    invisible, 4, 12–13, 15–16, 18, 20–23, 50, 61, 63, 71, 86, 103, 106, 121–122, 125, 127–128, 133, 137, 146, 158, 172, 183
    Four Worlds, 62, 98, 106, 108–116, 118, 121, 124–125, 137, 150, 177
    Great Chain of Being, 62, 64, 98, 106, 119, 125, 129, 137, 150, 159, 162, 177
    realms of spirit, 62, 96–99, 101, 103, 106, 108–109, 111, 112, 118, 122, 124–125, 127, 132, 133, 137, 158, 164–165, 172–173, 179, 182, 187, 191
    Spectrum of Consciousness, 46, 58, 65, 85, 98, 106, 118–120, 125, 137–138, 150, 154, 159, 177, 179, 183, 185–186, 191
non-ordinary
    altered states of consciousness, 27, 87

Index

non-ordinary—*Cont.*
  consciousness, 2–6, 11, 13, 16, 19–21, 23–25, 27, 29–31, 36–37, 39, 42–43, 46, 51–58, 60, 62, 64–65, 67, 73, 79, 85–87, 91–92, 95, 98–100, 103, 105–109, 111–112, 117–120, 122–127, 129–130, 135, 137–138, 140, 144–148, 150, 153–157, 159–160, 162–167, 172–173, 175, 177, 179–183, 185–189, 191–193
  mystical, 4, 15–16, 61–63, 66–67, 74, 91, 98, 102, 108–109, 113, 121, 136–137, 140, 146–147, 168, 185, 187
  states of awareness, 23, 53, 87, 107
Non-Self
  Buddhism, 37, 43, 53, 65, 102, 113, 120, 138, 180, 193
  Spectrum of Consciousness, 46, 58, 65, 85, 98, 106, 118–120, 125, 137–138, 150, 154, 159, 177, 179, 183, 185–186, 191
Now
  present, 5, 8–9, 11, 14, 26, 28, 32–33, 35–37, 57, 64, 85, 90, 99, 105, 108–109, 122, 127, 132, 134, 180, 186, 190–191
  time, 2, 4–5, 8, 10–15, 18–22, 24–30, 40–42, 51, 61, 70, 77, 83–84, 90–93, 101, 103, 107, 109–111, 117, 122–123, 125–128, 132–134, 144–146, 148–152, 155, 157, 159, 161–162, 165, 167–168, 173, 179–180, 182–187, 189, 192, 193
nowhen
  present, 5, 8–9, 11, 14, 26, 28, 32–33, 35–37, 57, 64, 85, 90, 99, 105, 108–109, 122, 127, 132, 134, 180, 186, 191–192
  time, 2, 4–5, 8, 10–15, 18–22, 24–32, 37, 40–42, 51, 61, 70, 77, 83–84, 90–93, 101, 103, 107, 109–111, 117, 122–123, 125–128, 132–134, 144–146, 148–152, 155, 157, 159, 161–162, 165, 167–168, 173, 179–180, 182–187, 189, 192, 193
nowhere
  present, 5, 8–9, 11, 14, 26, 28, 32–33, 35–37, 57, 64, 85, 90, 99, 105, 108–109, 122, 127, 132, 134, 180, 186, 191–192
  time, 2, 4–5, 8, 10–15, 18–22, 24–32, 40–42, 51, 61, 70, 77, 83–84, 90–93, 101, 103, 107, 109–111, 117, 122–123, 125–128, 132–134, 144–146, 148–152, 155, 157, 159, 161–162, 165, 167–168, 173, 179–180, 182–187, 189, 192, 193
Numinous
  ALL THAT IS, 35, 43, 98, 109, 111–112, 114–115, 119, 121, 127–128, 144, 147, 161–163, 166, 179, 189, 191
  SOURCE, 12, 15, 27, 38, 42, 64, 86–87, 99, 106, 108–109, 112–113, 115–116, 121–122, 125, 127, 134, 137, 148
  SPIRIT, 2, 11, 19, 61–64, 67, 86–87, 91, 93, 96–99, 101–109, 111–115, 118–120, 124–125, 128–129, 131, 133, 136–137, 139–141, 144, 147–148, 151, 157–159, 162, 164–166, 168–169, 174, 185, 193

Oath
  Childhood, 17, 28–29, 53, 92, 147, 153–154, 160, 180
  Communion, 29, 53, 58, 125, 147, 153–155, 177
  First People, 28–31, 33, 39, 40, 117–118, 133, 147, 153–154
  Great Mother, 54–56, 70, 92, 117, 125, 153–155, 159–160, 167
  Sanctuary, 153, 156
obedience
  Childhood, 17, 28–29, 53, 92, 147, 153–154, 160, 180
  Communion, 29, 53, 58, 125, 147, 153–155, 177
  First People, 28–31, 33, 39, 40, 117–118, 133, 147, 153–154
  Great Mother, 54–56, 70, 92, 117, 125, 153–155, 159–160, 167
  Sanctuary, 153, 156
objective
  see also Ego/Patriarchal Consciousness
  Patriarch, 54, 58, 73, 89, 91–92, 101, 125, 154–156, 159, 167, 175–177
  rational, 31–33, 38, 47, 50, 74–75, 99, 101, 103, 119, 155–156, 167, 192–193
  Separation, 28, 31, 43, 53–56, 58, 71, 76–78, 83, 86, 90, 101, 125, 153–155, 159, 167, 176–177, 190
observer
  participatory universe, 123, 127
  Quantum Mechanics, 13, 34, 35, 126, 183, 181, 186, 189, 192
Observer effect, 4
  participatory universe, 123, 127
  Quantum Mechanics, 13, 34, 35, 106, 126, 181, 183, 186, 189, 192
ocean of energy
  see also Bohm, D., 121
  Holomovement, 19–20, 37, 98, 106, 121–122, 124, 184, 185
Oedipal Complex
  Appendix III
  Crisis, 54–55, 73, 85, 154, 156, Appendix III
  Electra Complex, 55
  see also Ego/Patriarchal Consciousness
  Freud, S., 54–55, Appendix III
  Jung, C.G., 54, 66
  Neumann, E., 53–55, 65
  Patriarch, 54, 58, 73, 89, 91–92, 101, 125, 154–156, 159, 167, 175–177
  Separation, 28, 31, 43, 53–56, 58, 71, 76–78, 83, 86, 90, 101, 125, 153–155, 159, 167, 176–177, 190
  Steiner, R., 54–55
offspring
  offspring A-I, 113–124
  A-Jewish, 114–116
  B-Catholic, 114–116
  C-Yogi, 117
  D-Sufi, 117
  E-Shaman, 117–118
  F-Mayan, 118
  G-Huna, 118
  H-Hindu, 118–119
  I-Buddhism, 120
  Pathways, 3, 20–21, 23, 51, 91, 104, 113, 124–125, 157, 164
  Soul, 43, 45, 55, 58–64, 66, 85, 98, 102, 105, 111, 113–114, 119, 127, 132, 137, 148, 160, 164, 180, 189–191
  World of Creation, 64, 109
old paradigm
  see also Classical Science, 9, 28, 32–33, 81–82, 87, 180
  Science, 2–4, 6–15, 17, 19–29, 31–35, 37–41, 43–47, 50, 61–62, 64, 66, 76, 81–82, 86, 92–93, 98, 105, 107, 121, 127, 129, 139–141, 145, 155–156, 164, 177, 182, 187, 192–193, 167
  Separation, 28, 31, 43, 53–56, 58, 71, 76–78, 83, 86, 90, 101, 125, 153–155, 159, 167, 176–177, 190
Old Testament
  dreaming, 11, 110, 119–120, 126, 132–134, 146, 168, 180, 191
Old Testament—*Cont.*
  Genesis, 11, 60, 107–108, 121, 147–148, 164
  Visions, 52, 83, 91, 134, 137, 146–147, 158, 172 *see also* Mystics, Section six
One and the Many
  Godhead, 105, 109, 114, 183
  SOURCE, 12, 15, 27, 38, 42, 64, 86–87, 99, 106, 108–109, 112–113, 115–116, 121–122, 125, 127, 134, 137, 148
  SPIRIT, 2, 19, 61–64, 67, 86–87, 91, 93, 96–99, 101–109, 111–113, 115, 118–120, 124–125, 128–129, 131, 133, 136–137, 139–141, 144, 147–148, 151, 157–159, 162, 164–166, 168–169, 174, 185, 193
Ontogeny
  development of individual, 52–53, 54–56, Sections Four and Six
order
  Bohm, D., 13, 15–17, 19–20, 37, 98, 105–106, 121–126, 181, 183–184, 191
  Chaos, 9, 20, 28, 34, 37–38, 40, 42–43, 45–46, 52, 65, 84, 89, 101, 147, 156, 163–164, 173, 177, 180, 187, 79, 80, 159
  Complexity, 9, 20, 27, 34, 37, 42, 45–46, 52, 147, 150, 157, 159
  Explicate, 15–16, 20, 105, 121, 123, 124, 183–184, 191
  Four Worlds, 62, 98, 106, 108, 116, 118, 121, 124–125, 137, 150, 177
  Great Chain of Being, 62, 64, 98, 108, 119, 125, 129, 137, 150, 159, 162, 177
  Holomovement, 19–20, 37, 98, 106, 121–122, 124, 184
  Implicate, 13, 15–17, 20, 37, 45, 121–124, 126, 181, 183–185, 191
  levels, 2, 5, 15–16, 25, 29, 34, 57–64, 70, 85–86, 88, 94, 98–99, 102, 104–109, 111–112, 115–116, 118–125, 128, 130, 134–135, 137, 150, 157, 159, 163, 177, 181, 183–185, 188–190, 192 *see also* fields, ground
  time, 2, 4–5, 8, 10–15, 18, 20, 22, 24–32, 40–42, 47, 51, 54, 61, 77, 83–84, 90–93, 101, 103, 107, 109–111, 117, 122–123, 125–128, 132–134, 144–146, 148–151, 152, 155, 157, 159, 161–162, 165, 167–168, 173, 179–180, 182–187, 189–193
  Spectrum of Consciousness, 46, 58, 65, 85, 98, 106, 118–120, 125, 137–138, 150, 154, 159, 177, 179, 183, 185–186, 191
  Wilber, K., 62, 65–66
organizing principles
  archetypes, 22, 51–53, 59, 66, 88–89, 110, 122, 160, 163, 179, 191
  Evolution of Psyche, 118, 150, 154, 162, 165, 168
  Jung, C.G., 54, 66, 87–88
Origination
  Emanation, 62, 64, 105, 108–109, 114–116
  Four Worlds, 62, 98, 106, 108, 116, 118, 121, 124–125, 137, 150, 177
  Great Chain of Being, 62, 64, 98, 108, 119, 125, 129, 137, 150, 159, 162, 177
  Holomovement, 19–20, 37, 98, 106, 121–122, 124, 184
outer personality
  ASELF, 2, 70–94, 105, 110–112, 114, 126, 134–135, 138, 144–151, 153–166, 168, 173–174, 177, 182
  persona, 72–73, 88, 188
  SELF, 2–6, 8–9, 17–18, 22–24, 27–28, 31, 45, 50, 52–56, 58–59, 63–64, 66,

207

outer personality—*Cont.*
SELF—*Cont.*, 70, 72–74, 79–80, 86–93, 98–99, 101, 104, 106, 108–110, 112, 117–118, 125, 127–132, 134–139, 141, 144–147, 154–155, 158, 162, 165, 168, 185, 188–191
outer, lower sheath of annamayakosha
Buddhism, 37, 43, 67, 102, 120, 138, 180, 193
Hindu, 16, 20, 113, 119, 134, 186
Koshas, 113, 118–120, 185
offspring H, 118–120
Oxygen, 150, 188

pairs of electrons
non-locality, 13, 15–16, 18, 20, 23–24, 37, 123, 187
simultaneous time, 101, 125, 128, 167, 191
parallel
see also Multiple Worlds/realities
paranormal experiences
Altered States of Consciousness, 23, 27, 85–87, 91, 117, 125
non-ordinary experience, 16, 24, 41, 136
Parliament
Canada, 92, 145
Enough, 92–93, 144–145
parracide
Oedipus Crisis, 54, 55, 73, 154, Appendix III
Patriarch, 54, 58, 73, 89, 91–92, 101, 125, 154–156, 159, 167, 175–177
regression, 55, 160, 176
Separation, 28, 31, 43, 51, 53–56, 58, 71, 76–78, 81, 83, 86, 90, 94, 101, 125, 153–155, 159, 167, 176–177, 190
participatory Universe
Co-Creator, 162, 168
Observer Effect, 4, 7, 23, 36, 123
partnership, 110, 147, 162–163, 166–168
Transpersonal Human, 162, 173
particles
Personhood, 2–3, 5–6, 28, 50, 52, 54, 59, 65, 72, 75, 79, 85, 90, 109–111, 134, 136, 141, 154, 158, 162
Selves, 3–5, 25, 33, 50, 52, 59, 72–73, 75, 79, 81, 90, 92–93, 109, 127, 135, 144–145
shadow, 72–73, 75, 79, 88, 156, 159, 167, 176, 191
sub-personalities, 51–52, 73, 75, 78–79, 88
see also aspects and facets of personality
passive
Childhood, 17, 28–29, 53, 92, 125, 147, 153–154, 160
Communion, 29, 53, 58, 117, 125, 147, 153–155, 177
Pre-Personal Consciousness, 54, 58, 74, 92, 154, 168
path of Kabbalah
Jewish mysticism, 114–116
offspring A, 114–116
Path of the Heart
Four Worlds, 62, 98, 106, 108, 111–116, 118, 121, 124–125, 137, 150, 163, 177
Great Chain of Being, 62, 64, 98, 106, 119, 125, 129, 137, 150, 159, 162, 177
Realms of Spirit, 62, 93, 96–99, 101, 103–105, 107–109, 111, 113, 115, 117, 119, 121, 123–125, 127, 129, 131, 133, 135, 137, 139, 141, 144, 158, 162, 164–165
Spectrum of Consciousness, 46, 58, 65, 85, 98, 106, 118–120, 125, 137–138, 150, 154, 157, 159, 177, 179, 183, 185–186, 191

PATHWAYS
offspring A-I, 113–124
Patriarch
Choice, 24, 29, 42, 56, 65, 74, 78, 90, 111, 126, 149, 154, 159, 162, 166–167, 173
Ego-Consciousness, 54, 58, 73, 130, 154–155, 167
Individuation, 31, 54, 57, 77–78, 80–81, 88, 90, 95, 154, 185, 190
Masculine, 53–58, 73, 89, 99, 101, 125, 147, 160, 162–163, 172–173, 177, 190, 193
Oedipal Crisis, 54–55, 73, 154, Appendix III
Separation, 28, 31, 43, 51, 53–56, 58, 71, 76–78, 81, 83, 86, 90, 94, 101, 125, 153–155, 159, 167, 176–177, 190
patterns and blueprints
archetypes, 22, 51–53, 59, 66, 88–89, 110, 122, 160, 163, 179, 191
DNA, 18, 35, 88, 150, 181–183, 190
Morphogenetic forms 13, 21–22, 121–122, 186 *see also* forms
World of Formation, 64, 98, 108–111, 113, 126, 131
perceptions
awareness, 2, 4, 6, 20, 23, 27–28, 30, 39, 53–58, 60, 85, 87, 90, 92, 98, 99, 101, 105–107, 109, 112–113, 117–122, 125, 127, 130–131, 133, 137, 145, 147, 155–157, 159, 162–163, 166, 168, 176, 180, 184–186, 190
Masculine & Feminine, 52–55, 193
Perennial Philosophy
Huxley, A., 85–86, 105, 188
mystics, 3, 12–13, 16, 19–20, 27, 41, 79, 86, 98, 102, 107, 114, 117, 121, 125, 127, 130, 138–139, 164, 187–188
universal, 9, 54, 70, 86, 183, 187
Wilber, K., 62, 65–66
personal
evolution, 1, 3, 5–6, 9, 53, 55, 57, 62, 67, 88, 91, 93, 95, 98–99, 104–107, 112–114, 118, 122, 125, 127, 129, 135, 137, 139–140, 147–148, 150–152, 154, 159–160, 162, 165–168, 173, 177, 182, 185–186, 188, 190–191
Explicate, 15–16, 20, 105, 121, 123, 124, 183–184, 191
ontogeny, 52–53, 151–154, 165, 167–168, 187–188, 189
persona, 72–73, 88, 188
personal unconscious, 4, 51–52, 59, 73, 75, 79–80, 88, 185, 188
Self, 2–6, 8–9, 17–18, 22–24, 27–28, 31, 45, 50, 52, 56, 58–59, 63–64, 66, 70, 72–74, 79–80, 86–93, 98–99, 101, 104, 106, 108–110, 112, 117–118, 125, 127–132, 134–139, 141, 144–147, 154–155, 158, 162, 165, 168, 185, 188–191
World of Manifestation, 64, 98, 105, 111–112, 113–116, 121–124, 124, 130, 147, 150, 155
photosynthesis
evolution, 1, 3, 5–6, 9, 53, 55, 57, 62, 67, 88, 91, 93, 95, 98–99, 104–107, 112–114, 118, 122, 125, 127, 129, 135, 137, 139–140, 147–148, 150–152, 154, 159–160, 162, 165–168, 173, 177, 182, 185–186, 188, 190–191
oxygen, 150, 188
prokaryote, 149–150, 152
phylogeny
development and evolution of species, 52–54, 151–154, 165, 167–168, 188
physical
Buddhism, 37, 43, 65, 102, 113, 120, 138, 180, 193

physical—*Cont.*
density, 111–112, 128, *see also* physicality
Explicate order, 15–16, 123, 183–184, 191
illusion, 16, 18, 36, 130, 186
virtual reality, 130–131, 135, 139, 181, 192
World of Manifestation, 64, 98, 105, 111–112, 113–116, 121–124, 124, 130, 147, 150, 155
Plato
Cave, 172–173
World of 2070 A.D., 168, 172–173
Popul Vuh
Mayan, 40, 113, 118
potentia
entelechy, 51, 182
Plato, 14, 51, 173–174
powerful God
Classical, 9, 28, 32, 152, 180, 184–185, 187–189, 191
Old Paradigm Science, 9–10, 32, 33, 81, 167
pranamayokosha
Buddhism, 37, 102, 120, 180, 193
Hindu, 16, 20, 113, 119, 134, 186
Kosha, 113, 118–120, 185
offspring H, 118–120
pre-conscious
birth, 9–10, 12, 20, 26–27, 32, 41, 54, 57, 70, 95, 105, 107–109, 111, 123–124, 128–129, 132, 147, 153–154, 159, 161–162, 164, 180, 189
Childhood, 17, 28–29, 53, 92, 125, 147, 153–154, 160, 177
Communion, 29, 53, 58, 125, 147, 153–155, 177
Goddess, 53–54, 56–58, 66–67, 70, 92, 147, 154–155, 160, 167, 169, 175–177
Great Mother, 54–56, 70, 92, 117, 125, 153–155, 159–160, 167
Matriarch, 53, 56–58, 154
pre-personal
Childhood, 17, 28–29, 53, 92, 125, 147, 153–154, 160, 177
Communion, 29, 53, 58, 125, 147, 153–155, 177
Goddess, 53–54, 56–58, 66–67, 70, 92, 147, 154–155, 160, 167, 169, 175–177
Great Mother, 54–56, 70, 92, 117, 125, 153–155, 159–160, 167
Matriarch, 53, 56–58, 154
pre-temporal
peri-natal, 20
predictable
Classical, 9, 28, 32, 152, 180, 184–185, 187–189, 191
linear, 10, 26, 31–32, 38, 40, 125, 155, 157, 167, 187
Newtonian, 32–34, 182, 186, 189, 192
Old Paradigm Science, 9–10, 31, 32, 33, 81, 167
Pribram, Carl
holograms, 18
memory, 18
Prigogine, Ilya
chaos, 9, 20, 28, 34, 37–38, 40, 42–43, 45–46, 52, 65, 84, 88–89, 101, 147, 156, 163–164, 173, 177, 180, 187
complexity, 9, 20, 27, 34, 37, 42, 45–46, 52, 147, 150, 157, 159
dissipative structures, 9, 20, 28, 34
primordial bacteria
evolution, 1, 3, 5–6, 9, 53, 55, 57, 62, 67, 88, 91, 93, 95, 98–99, 104–107, 112–114, 118, 122, 125, 127, 129, 135, 137, 139–140, 147–148, 150–152, 154, 159–160, 162, 165–168, 173, 177, 182, 185–186, 188, 190–191
prokaryote, 149–150, 152
uni-cellular, 150

Index

process of transformation and enlightenment
   evolution, 1, 3, 5–6, 9, 53, 55, 57, 62, 67, 88, 91, 93, 95, 98–99, 104–107, 112–114, 118, 122, 125, 127, 129, 135, 137, 139–140, 147–148, 150–152, 154, 159–160, 162, 165–168, 173, 177, 182, 185–186, 188, 190–191
   Path of the Heart, 93, 98–99, 103–104, 124, 127, 137, 144, 157, 160, 162, 165
   Perennial Psychology, 85–86, 105, 188
   Spectrum of Consciousness, 46, 58, 65, 85, 98, 106, 118–120, 125, 137–138, 150, 154, 159, 177, 179, 183, 185–186, 191
   Transpersonal Psychology, 58, 59, 62–64, 85, 87, 94, 118, 162–168, 177, 192
   Wilber, K., 62, 65–66
projections
   ego, 22, 55–56, 58–59, 64, 72–73, 86, 88, 90, 93, 98, 112–113, 118, 125, 128, 130, 134, 136, 140, 145, 147, 154–155, 167, 177, 182, 184, 191 - *see also* Ego Consciousness
   heal split 58, 64, 87, 88, 176
   reflection, 19, 43, 111–112, 186
   responsibility, 6, 55, 65, 90, 118, 127, 129, 133, 145–147, 154, 159–160, 165, 176, 190
   shadow, 72–73, 75, 79, 88, 156, 159, 167, 176, 191
Prokaryote
   evolution, 1–3, 5–6, 9, 53, 55, 57, 62, 67, 88, 91, 93, 95, 98–99, 104–107, 112–114, 118, 122, 125, 127, 129, 135, 137, 139–140, 147–148, 150–152, 154, 159–160, 162, 164–168, 173, 177, 182, 185–186, 188, 190–191
   extinct, 39, 152
   oxygen, 150, 188
   photosynthesis, 150–152, 188
   uni-cellular, 150
Prophets
   Seer, 86–87, 117, 190
   Visionaries, 130, 146, 168, 172
   World of 2070 A.D., 168, 172
psyche
   evolution, 1, 3, 5–6, 9, 53, 55, 57, 62, 67, 88, 91, 93, 95, 98–99, 104–107, 112–114, 118, 122, 125, 127, 129, 135, 137, 139–140, 147–148, 150–152, 154, 159–160, 162, 165–168, 173, 177, 182, 185–186, 188, 190–191
   Four Worlds, 62, 98, 106, 108–116, 118, 121, 124–125, 137, 150, 177
   Great Chain of Being, 62, 64, 98, 106, 119, 125, 129, 137, 150, 159, 162, 177
   Spectrum of Consciousness, 46, 58, 65, 85, 98, 106, 118–120, 125, 137–138, 150, 154, 159, 177, 179, 183, 185–186, 191
psycho-social development
   archetypes, 22, 51–53, 59, 66, 88–89, 110, 122, 160, 163, 179, 191
   evolution of the psyche, 1, 3, 5, 53, 62, 88, 98–99, 101, 105–106, 113, 125, 135, 137
   Spectrum of Consciousness, 46, 58, 65, 85, 98, 106, 118–120, 125, 137–138, 150, 154, 159, 177, 179, 183, 185–186, 191
psychological
   Electra Complex, 55
   see also Ego/Patriarchal Consciousness
   Oedipus Crisis, 54–55, 73, 85, 154, 156, Appendix III
   Patriarch, 54, 58, 73, 89, 91–92, 101, 125, 154–156, 159, 167, 175–177
   Separation, 28, 31, 43, 53–56, 58, 71, 76–78, 81, 83, 86, 90, 101, 125, 153–155, 159, 167, 176–177, 190

Psychology
   Schools, 58–59, 62
   Section three & four, pp. 49–94
   Third/Fourth Force, 58–59, 62–63, *see also* Transpersonal Psychology
psychotherapy
   ANNIE, 3, 79, 84–89, 91
   ASELF, 2, 70–94, 105, 110–112, 114, 126, 134–135, 138, 144–151, 153–166, 168, 173–174, 177, 182
Puer
   archetypes, 22, 51–53, 59, 66, 88–89, 110, 122, 160, 163, 179, 191
   Collective Unconscious, 5–6, 51, 59, 63, 88, 91, 179, 181
   Jung, C.G., 66, 87–88
quantum mechanics
   Bell's Theorem, 15
   see also Bohm, David
   Holomovement, 19–20, 37, 98, 106, 121–122, 124, 184
   non-locality, 13, 15–16, 18, 20, 23–24, 37, 121–123, 187
   Observer Effect, 4, 23, 36, 43, 123, 127
   physics, 4, 9, 15–16, 23, 34–35, 43–46, 62, 76, 106, 121, 139, 152, 180–181, 186, 189, 192
   potential, 3–5, 12, 16, 18, 20, 29, 32, 36, 43, 53, 59–60, 65, 85, 87–88, 108–109, 117, 122, 124, 126, 128, 131–132, 134, 145, 147–148, 162, 172–173, 183, 189, 192
   Uncertainty, 2, 4, 9, 19, 28, 34, 36–38, 42–43, 84, 147, 156, 159, 163, 167, 173, 177, 192 - *see also* Chaos, Heisenberg
quarks, 4, 34, 35, 44–46, 189
   6 flavors, leptons, 35, 126, 183

rational
   Classical, 9, 28, 32, 152, 180, 184–185, 187–189, 191
   Ego-Consciousness, 73
   logical, 31, 50, 75, 155, 167
   Old Paradigm Science, 9–10, 32, 33, 81, 167
   Patriarch, 54, 58, 73, 89, 91–92, 101, 125, 154–156, 159, 167, 175–177
   Separation, 28, 31, 43, 53–56, 58, 71, 76–78, 83, 86, 90, 101, 125, 153–155, 159, 167, 176–177, 190
   RE-CONCILIATION AND UNION *see also* pages 28, 40, Appendix III
   levels, 2, 5, 15–16, 25, 29, 34, 57–64, 70, 85–86, 88, 94, 98–99, 102, 104–109, 111–112, 115–116, 118–125, 128, 130, sacred marriage, 3, 57–59, 89, 101, 125, 138, 147, 163, 168, 172, 177, 190
   transpersonal, 57–59, 62–64, 85, 87–88, 91–93, 98, 101, 112–113, 118, 125, 152, 160, 162–168, 173, 177, 190, 192 134–135, 137, 150, 157, 159, 163, 177, 181, 183–185, 188–190, 192
real world
   earth, 2–3, 8–11, 20, 22, 26, 28–31, 33, 35, 38–42, 45, 56–57, 65, 70, 92, 94, 101, 105–106, 108, 110, 113, 116–118, 121, 127–128, 135–136, 145, 147–153, 155–157, 161–164, 166, 169, 173, 175–176, 184, 188 - *see also* Mother Earth
   physicality, 2, 15, 41, 90, 109, 111, 124, 126, 130–131, 135, 137, 148, 153, 157, 159, 163–164, 186
   World of Manifestation, 64, 98, 105, 111–112, 113–116, 121–124, 124, 130, 147, 150, 155
Reality Principle - 99–101
   see also Ego-Consciousness

Reality Principle—*Cont.*
   Classical, 9, 28, 32, 152, 180, 184–185, 187–189, 191
   Old Paradigm Science, 9–10, 32, 33, 81, 167
   Separation, 28, 31, 43, 53–56, 58, 71, 76–78, 83, 86, 90, 101, 125, 153–155, 159, 167, 176–177, 190
realm
   domains, 59, 79, 96, 122, 186
   Four Worlds, 62, 98, 106, 108, 118, 121, 124–125, 137, 150, 177
   Great Chain of Being, 62, 64, 98, 108, 119, 125, 129, 137, 150, 159, 162, 177
   levels, 2, 5, 15–16, 25, 29, 34, 57–64, 70, 85–86, 88, 94, 98–99, 102, 104–109, 111–112, 115–116, 118–125, 128, 130, 134–135, 137, 150, 157, 159, 163, 177, 181, 183–185, 188–190, 192
   spectrum, 15, 36, 46, 58, 65, 85–86, 98, 106, 108, 112, 118–120, 122, 125, 131, 137–138, 150, 154, 159, 177, 179, 181, 183, 185–186, 191 *see also* Spectrum of Consciousness
receptive
   Feminine principle, 53–56, 101, 160
   intuitive, 50, 53, 99, 101, 160, 167, 190
   mode of perception, 99, 101, 156
   relational, 91, 93, 167
   transpersonal, 57–59, 62–64, 85, 87–88, 91–93, 98, 101, 112–113, 118, 125, 152, 160, 162–168, 173, 177, 190, 192,
reclaim ASOUL
   heal the split, 58, 64, 87–88, 176
   integrity, 90, 112, 159, 163
   wholeness, 2–3, 16, 18, 45, 53, 56–57, 60, 88, 91, 99, 112, 118, 122, 129, 137, 145, 156, 159–160, 163, 189–190, 193
redemption
   Feminine, 53–58, 66, 74–75, 89, 91–94, 99, 101, 117, 125, 147, 155, 159–160, 162–163, 167, 172–173, 175–177, 185, 190, 193
   Transpersonal Human, 162, 173
   Wholeness, 2–3, 16, 18, 45, 53, 56–57, 60, 88, 91, 99, 112, 118, 122, 129, 137, 145, 156, 159–160, 163, 189–190, 193
reincarnation
   creation & annihilation, 41, 42, 164
   cycle of birth & death, 42, 162, 164, 189
   Karma, 37, 128–129, 185
Relational Self
   Feminine, 53–58, 66, 74–75, 89, 91–94, 99, 101, 117, 125, 147, 155, 159–160, 162–163, 167, 172–173, 175–177, 185, 190, 193
   Transpersonal, 57–59, 62–64, 85, 87–88, 91–93, 98, 101, 112–113, 118, 125, 152, 160, 162–168, 173, 177, 190, 192
religions
   Pathways A-I, 113–124
repertoire
   blueprints, 13, 18, 21, 35, 51–52, 88, 110–112, 122, 131, 133, 150, 181–183, 190
   patterns, 21, 23, 30, 37, 41, 45, 51–52, 59, 61, 79–80, 84–85, 89, 110, 121, 131, 140, 161, 179, 183, 188, *see also* forms
   scripts, 122, 133
Repression
   banished, 50, 75, 156, 159
   dark side, 145, 156
   disowned, 51–52, 55, 58, 72–73, 75, 79, 88, 145, 156, 185
   shadow, 72–73, 75, 79, 88, 156, 159, 167, 176, 191
resolution, 3, 34, 54–55, 73, 117, 154, 187

209

responsibility
  adulthood, 54, 144, 160, 173
  Feminine, 53–58, 66, 74–75, 89, 91–94,
    99, 101, 117, 125, 147, 155, 159–160,
    162–163, 167, 172–173, 175–177, 185,
    190, 193 see also Emerging Feminine
  Transpersonal, 57–59, 62–64, 85, 87–88,
    91–93, 98, 101, 112–113, 118, 125,
    152, 160, 162–168, 173, 177, 190, 192
resurrector
  Feminine Principle, 52–56, 101, 160
revelation
  Alexandra, 135–136
  gnosis, 87, 129, 136–137, 183
  Jesus, 60, 114, 116, 135–136, 138
  Mary, 135–136
Rising Star, 82–85, 89
ritual of initiation
  Electra Complex, 55
  Oedipal Crisis, 54, 55, 73, 154, Appendix
    III
Rowan, J.
  Inner Selves, 4, 52, 72–73, 79, 92, 144
  sub-personalities, 51–52, 73, 75, 78–79,
    88
Ruah
  Jewish Mysticism, 114
  offspring A, 114
  Soul level, 114

sacred
  act of creation, 86, 99, 126, 133, 147, 155
  ALL THAT IS, 35, 43, 98, 109, 111–112,
    114, 119, 121, 127–128, 144, 147,
    161–163, 166, 179, 189, 191
  Sacred Covenant, 153, 155
  Sacred Marriage, 3, 57–59, 89, 101, 125,
    138, 147, 163, 168, 172, 177, 190
  sacred vessels, 165–166
safety
  Childhood, 17, 28–29, 53, 92, 125, 147,
    153–154, 160, 177
  Communion, 29, 53, 58, 125, 147,
    153–155, 177
  First People, 28–31, 33, 39, 40, 117–118,
    133, 147, 153–154
  Mother Earth, 3, 12, 28–30, 39–40,
    52–53, 56, 70, 92, 149, 153, 155, 161,
    167, 173
  sanctuary, 153, 156
scapegoat
  bad father, 156
  bad mother, 155, see also Section 3, Slaying The Dragon
  Science, 2–4, 6–15, 19–23, 26–28, 31–37,
    41, 43, 50, 61–62, 64, 76, 81–82, 98,
    121–122, 126–127, 155–156, 164, 177,
    182, 187, 192–193
Science
  new paradigm, 9, 34, 37–39, 50, 59, 86,
    121, 138, 167
  Old Paradigm, 9–10, 32, 33, 65, 81, 167
Scientist
  ASELF, 2, 70–94, 105, 110–112, 114,
    126, 134–135, 138, 144–151, 153–166,
    168, 173–174, 177, 182
second epoch
  see also Ego-Consciousness
  Patriarch, 54, 58, 73, 89, 91–92, 101, 125,
    154–156, 159, 167, 175–177
  Separation, 28, 31, 43, 53–56, 58, 71,
    76–78, 83, 86, 90, 101, 125, 153–155,
    159, 167, 176–177, 190
Sefirot
  Jewish Mysticism, 113, 114, 116
  Kabbalah, 113–116, 185
  offspring A, 114

Self, 2–6, 8–9, 17–18, 22–24, 27–28, 31, 45,
  50, 52–56, 58–59, 63–64, 66, 70, 72–74,
  79–80, 86–93, 98–99, 101, 104, 106,
  108–110, 112, 117–118, 125, 127–132,
  134–139, 141, 144, 147, 154–155, 158,
  162, 165, 168, 185, 188–191
self-organizing properties
  Chaos, 9, 20, 28, 34, 37–38, 40, 42–43,
    45–46, 52, 65, 79, 80, 84, 89, 101, 147,
    156, 159, 163–164, 173, 177, 180, 187
  Complexity, 9, 20, 27, 34, 37, 42, 45–46,
    52, 147, 150, 157, 159
  non-linear, 9, 20, 27, 32, 36–38, 40, 42,
    62, 86, 180, 189, 193, 147, 150, 182,
    184, 187, 190–191
  Prigogine, I., 20, 28, 34, 38, 181–182
SELFHOOD
  evolution, 1, 3, 5–6, 9, 53, 55, 57, 62, 67,
    88, 91, 93, 95, 98–99, 104–107,
    112–114, 118, 122, 125, 127, 129, 135,
    137, 139–140, 147–148, 150–152, 154,
    159–160, 162, 165–168, 173, 177, 182,
    185–186, 188, 190–191
  Four Worlds, 62, 98, 106, 108–116, 118,
    121, 124–125, 137, 150, 177
  Great Chain of Being, 62, 64, 98, 106,
    119, 125, 129, 137, 150, 159, 162, 177
  Spectrum of Consciousness, 46, 58, 65,
    85, 98, 106, 118–120, 125, 137–138,
    150, 154, 159, 177, 179, 183, 185–186,
    191
sensory apparatus
  brain, 13, 17–18, 32, 37, 61, 87, 95, 99,
    169
  neurological apparatus, 13, 25, 125
Separation
  archetype, 53–54, 88, 153, 179
  see also Ego-Consciousness
  see also Jung, C.G.
  Patriarch, 54, 58, 73, 89, 91–92, 101, 125,
    154–156, 159, 167, 175–177
  Separation, 28, 31, 43, 53–56, 58, 71,
    76–78, 83, 86, 90, 101, 125, 153–155,
    159, 167, 176–177, 190
  separation and individuation, 31, 54,
    77–78, 81, 90, 154
  separation anxiety, 76, 94, 190
seven chakras
  Kundalini, 117, 140, 185, 187
  offspring C, 117
seven inner worlds
  offspring D, 117
  Sufi, 117
Shadow
  banished, 50, 75, 156, 159
  disowned, 51–52, 55, 58, 72–73, 75, 79,
    88, 145, 156, 185
  inner children, 51–52, 75, 88, 145
  inner selves, 4, 52, 72–73, 79, 92, 144
  Jung, C.G., 66, 87–88
  split-off, 72
Shaman, offsprings E & F, 117–118
Sheldrake
  blueprints, 21, 51–52, 110–112, 122, 131,
    133, see also DNA
  forms, 6, 11, 13, 22–23, 28, 30–31, 38,
    40–42, 51, 77, 79, 85–88, 105–110,
    118, 121–122, 124, 129, 135, 150–151,
    179, 188, see also patterns
  least action pathway, 21–23, 185
  Morphogenetic Fields, 13, 21–22,
    121–122, 135, 187
shift in Consciousness
  non-ordinary, 16–17, 23–24, 85–87, 91,
    125, 132–133
  states of consciousness, 11, 23, 27,
    85–87, 91, 120, 125, 134, 180, 192
silence, 78, 87, 91

Simultaneous
  Karma, 128–129, 185
  non-linear, 20, 27, 37, 42, 86, 184, 187,
    190–191
  reincarnation, 37, 128–129, 189
  time, 2, 4–5, 8, 10–15, 18–22, 24–32,
    40–42, 51, 61, 70–77, 83–84, 90–93,
    101, 103, 107, 109–111, 117, 122–123,
    125–128, 132–134, 144–146, 148–152,
    155, 157, 159, 161–162, 165, 167–168,
    179–180, 182–187, 189, 192–193
singularity
  Big Bang, 10, 12–14, 20, 32, 180
SLAY THE DRAGON
  Goddess, 53–54, 56–58, 66–67, 70,
    92, 147, 154–155, 160, 167, 169,
    175–177
  Great Mother, 54–56, 70, 92, 117, 125,
    153–155, 159–160, 167
  Oedipal Complex, 54–55, 73, 154,
    Appendix III
  Patriarch, 54, 58, 73, 89, 91–92, 101, 125,
    154–156, 159, 167, 175–177
  Separation, 28, 31, 43, 53–56, 58, 71,
    76–78, 81, 83, 86, 90, 101, 125,
    153–155, 159, 167, 176–177, 190
socialization
  ASELF, 2, 70–94, 105, 110–112, 114,
    126, 134–135, 138, 144–151, 153–166,
    168, 173–174, 177, 182
  Childhood, 17, 28–29, 53, 92, 125, 147,
    153–154, 160, 177
software disk
  illusion, 16, 18, 36, 130, 186
  Joe, 130–131
  MAYA, 16, 110, 130, 186
  virtual reality, 16, 130–131, 135, 139,
    181, 192
solar disk
  evolution, 1, 3, 5–6, 9, 53, 55, 57, 62, 67,
    88, 91, 93, 95, 98–99, 104–107,
    112–114, 118, 122, 125, 127, 129, 135,
    137, 139–140, 147–148, 150–152, 154,
    159–160, 162, 165–168, 173, 177, 182,
    185–186, 188, 190–191
solar system, 25, 149, 152
Sorcerer
  Art of Dreaming, 132–133, 168, 191
  Bardos, 119, 120, 133, 180
  Castaneda, C., 66, 132–133
  lucid dreaming, 133
  Wolf, F., 23, 44, 67, 121, 126, 133, 140
Soul
  ASOUL, 2–5, 108–110, 112–114,
    118–120, 124, 126–128, 131–135,
    138, 160, 164, 168, 172, 174
  Super-Implicate, 121
  World of Creation, 64, 109
Source
  ALL THAT IS, 35, 43, 98, 102, 108–109,
    111–112, 119, 121, 127–128, 144, 147,
    161–163, 166, 179, 183, 189, 191
  SOURCE, 12, 15, 27, 38, 42, 64, 86–87,
    99, 106, 108–109, 112–113, 115–116,
    121–122, 125, 127, 134, 137, 148
  SPIRIT, 2–3, 19, 61–64, 67, 86–87, 91,
    93, 96–99, 101–105, 107–109,
    111–115, 117–120, 124–125, 127–128,
    131, 133, 136–137, 139–141, 144,
    147–148, 151, 157–159, 162, 164–166,
    168–169, 174, 185, 193
  World of Origination, 62, 64, 98,
    108–109, 113–116, 118, 121–122,
    124
space-time
  World of Manifestation, 42, 62, 64, 98,
    105, 108, 111–112, 118, 121–122, 124,
    130, 147, 150, 155

210

Index

spark of Divinity
  Soul, 43, 45, 55, 58–64, 66, 85, 98, 102, 105–111, 113–114, 119, 127, 132, 137, 148, 160, 164, 180, 189–191
specialization
  see also Classical Science
  see also Ego-Consciousness
  Patriarch, 54, 58, 73, 89, 91–92, 101, 125, 154–156, 159, 167, 175–177
  Separation, 28, 31, 43, 51, 53–56, 58, 71, 76–78, 81, 83, 86, 90, 94, 101, 125, 153–155, 159, 167, 176–177, 190
Spectrum of Consciousness
  Transpersonal Psychology, 58–59, 62–64, 85, 87, 94, 118, 162–168, 177, 192, 194
  Wilber, K., 62, 65–66
speed of light
  Einstein, A., 9, 15, 21, 35, 186, 189
  electrons, 15, 21, 123, 179, 186 see also sub-atomic particles
  non-locality, 13, 15–16, 18, 20, 23–24, 37, 121–123, 187
spiral journey
  Path of the Heart, 93, 98–99, 103–104, 124, 127, 137, 144, 157, 160, 162, 165
  Realms of Spirit, 62, 96–99, 101, 103, 106, 108–109, 111, 112, 118, 122, 127, 132, 133, 137, 158, 164–165, 172–173, 179, 182, 187, 191
  Spectrum of Consciousness, 46, 58, 65, 85, 98, 106, 118–120, 125, 137–138, 150, 154, 159, 177, 179, 185–186, 191
spirit
  ALL THAT IS, 35, 43, 70, 98, 109, 111–112, 114, 119, 121, 127–128, 144, 147, 161–163, 166, 179, 183, 189, 191
  SOURCE, 12, 15, 18–19, 27, 35, 38, 42, 63–64, 86–87, 99, 106, 108–109, 112–113, 115–116, 119–122, 125, 127, 131, 134, 137, 148, 152, 158, 179, 183, 193
  SPIRIT, 2, 11, 19, 61–64, 67, 86–87, 91, 93, 96–99, 101–109, 111–115, 118–120, 124–125, 128–129, 131, 133, 136–137, 139–141, 144, 147–148, 151, 157–159, 162, 164–166, 168–169, 174, 185, 193
spiritual and sacred science, 92
Spirituality
  crisis, 29, 34, 38, 54–55, 73, 85, 154, 156
  enlightenment, 59, 62, 64, 98, 103–104, 106–107, 125, 128, 182, 189
  heritage, 40, 103, 106, 130, 161
  journey, 2, 5, 26, 55, 57–58, 61, 70, 85–88, 93, 98, 104, 110, 112, 117, 124–125, 127, 133–134, 136–139, 151, 159–160, 164–165, 168, 181
  pathways, 3, 20–21, 23, 51, 91, 104, 113, 124–125, 157, 164
  Spiritual birth, 57, 147, 161–162, 163
  spiritual traditions, 3, 10, 13–14, 27, 39, 61–62, 64, 85–86, 94, 99, 105, 107–108, 128, 130, 177
  transformation, 23, 57–58, 60, 85, 88, 92, 103, 106, 125, 128, 145, 156–157, 162, 186, 192
  unfolding, 2–3, 28, 75, 88, 106, 117, 122, 155, 164, 184, 186
split-off
  heal the split, 58, 64, 87–88, 176
  inner selves, 4, 52, 72–73, 79, 92, 144, 155, see also aspects and facets of personality, archetypes
  Path of the Heart, 59, 93, 98–99, 103–104, 124, 127, 137, 144, 157, 160, 162, 165
  sub-personalities, 51–52, 73, 75, 78–79, 88
  shadow, 72–73, 75, 79, 88, 156, 159, 167, 176, 191

split-off—Cont.
  transformation, 23, 57–58, 60, 85, 88, 92, 103, 106, 125, 128, 156–162, 186, 192
St. Augustine, 32
states
  after-death, 120, 133, 180
  Bardos, 37, 120, 133, 180
  Buddhism, 37, 43, 67, 102, 120, 138, 180, 193
  dreaming, 11, 110, 119–120, 126, 132–134, 146, 168, 180, 191
  meditation, 16, 27, 41, 91, 120, 180–181, 186
  re-birth, 120, 180, 189
Steiner, Rudolph
  Oedipal Crisis, 55, 154, Appendix III
Stop the world
  Mayan, 40, 113, 118
  offspring F, 118
  sorcerer, 118, 133, 191
storehouse
  implicate, 13, 15–17, 20, 37, 45, 121–124, 126, 181, 183–185, 191
  World of Formation, 62, 64, 98, 110–111, 113, 121–122, 126, 131
sub-atomic
  Quantum Mechanics, 13, 34, 181, 186, 189, 192
sub-personalities
  inner selves, 4, 52, 72–73, 79, 92, 144
  multi-faceted, 2–3, 5, 8, 18, 52, 59, 61, 70–72, 75, 79, 81, 88, 90, 92, 105, 119, 156, 166–167, 190
  personal unconscious, 4, 51–52, 59, 73, 75, 79–80, 88, 185, 188
  shadow, 72–73, 75, 79, 88, 156, 159, 167, 176, 191
  Rowan, J., see also sub personalities
subjective
  Feminine, 53–58, 74–75, 89, 91–94, 99, 101, 117, 125, 147, 155, 159–160, 162–163, 167, 172–173, 175–177, 185, 190, 193
  inner, 2, 4, 12–13, 19, 25, 27, 29, 33, 41, 51–52, 55–57, 59–60, 65, 67, 70, 72–73, 75–76, 78–82, 86–88, 91–92, 99, 102–104, 110, 112, 115, 117, 122, 124, 132, 134, 144–145, 147, 155, 161, 163–165, 167, 172–173, 183, 185, 192
  intuitive, 50, 53, 99, 101, 153, 160, 167
  New Paradigm, 9, 34, 37–39, 50, 59, 86, 121, 138, 167
Sufi mystic
  alam-al-mithal, 117
  offspring D, 117
SUPER IMPLICATE
  See also Bohm, D.,
  Four Worlds, 62, 98, 106, 108, 118, 121, 124–125, 137, 150, 177
  Holomovement, 19–20, 37, 98, 106, 121–122, 124, 184, 191
  World of Creation, 64, 109, 98, 121–122
  World of Formation, 64, 98, 110–112, 121–122, 124, 126, 131
Superego
  Freud, S., 54, 72–73, 92–93, 134, 191
supernova
  evolution, 1, 3, 5–6, 9, 53, 55, 57, 62, 67, 88, 91, 93, 95, 98–99, 104–107, 112–114, 118, 122, 125, 127, 129, 135, 137, 139–140, 147–148, 150–152, 154, 159–160, 162, 165–168, 173, 177, 182, 185–186, 188, 190–191
  universe, 2, 4, 6, 9–10, 12–21, 23–25, 28–29, 31–32, 34–35, 40–43, 50, 61, 64–65, 85–86, 88, 92–93, 106–107, 109–110, 112, 114, 123, 126–127, 129–130, 132–135, 137, 147–149, 151,

supernova
  universe—Cont., 153, 155, 159, 161, 163–166, 168, 172, 177, 179–182, 186–187, 189, 191–193
survival of Humankind
  crossroads, 164, 166
  Destiny, 42, 136, 157, 159, 166, 173, 179
  evolution, 1, 3, 5–6, 9, 53, 55, 57, 62, 67, 88, 91, 93, 95, 98–99, 104–107, 112–114, 118, 122, 125, 127, 129, 135, 137, 139–140, 147–148, 150–152, 154, 159–160, 162, 165–168, 173, 177, 182, 185–186, 188, 190–191
Symbiosis
  Childhood, 17, 28–29, 53, 92, 125, 147, 153–154, 160, 177
  Communion, 29, 53, 58, 125, 147, 153–155, 177
  Goddess, 53–54, 56–58, 66–67, 70, 92, 147, 154–155, 160, 167, 169, 175–177
  Matriarch, 53, 56–58, 154
  Mother, 14, 20, 28–30, 33, 39–40, 51, 53–56, 59, 70–74, 76, 78–81, 83–84, 90–95, 117, 125, 136, 145, 149, 153–155, 159–161, 167, 173, 176–177, 180, 190–192

Tao Te Ching
  Buddhism, 37, 43, 65, 102, 113, 120, 138, 180, 193
  genesis, 11, 60, 107–108, 121, 147–148, 164
technology
  Old Paradigm & Classical Science, 9–10, 28, 32–33, 81, 152, 180, 184–185, 187–189, 191
  Patriarch, 54, 58, 73, 89, 91–92, 101, 125, 154–156, 159, 167, 175–177
ten energy centers
  Kabbalah, 113–116, 185
  offspring A, 114
  sefirot, 114
THE ART OF DREAMING
  Bardos, 119,120, 133, 180
  Castaneda, C., 66, 132–133
  LaBerge, S., 133
  lucid dreaming, 133
  Wolf, F., 23, 44, 67, 121, 126, 133, 140
The First People
  Childhood, 17, 28–29, 53, 92, 125, 147, 153–154, 160, 177
  Communion, 29, 53, 58, 125, 147, 153–155, 177
  indigenous, 29–30, 92
The Observer Effect
  Explicate Order, 15–16, 123, 183–184, 191
  Heisenberg, W., 36 - see also Chaos and Uncertainty
  Quantum Mechanics, 4, 13, 15, 16, 34, 181, 186, 189, 192
The One and the Many
  ALL THAT IS, 35, 43, 98, 109, 111–112, 114, 119, 121, 127–128, 144, 147, 161–163, 166, 179, 189, 191
  SOURCE, 12, 15, 27, 38, 42, 64, 86–87, 99, 106, 108–109, 112–113, 115–116, 121–122, 125, 127, 134, 137, 148
  SPIRIT, 2–3, 19, 61–64, 67, 86–87, 91, 93, 96–99, 101–105, 107–109, 111–115, 117–120, 124–125, 127–128, 131, 133, 136–137, 139–141, 144, 147–148, 151, 157–159, 162, 164–166, 168–169, 174, 185, 193
The Path of the Heart
  Four Worlds, 62, 98, 106, 108–116, 118, 121, 124–125, 137, 150, 177
  Great Chain of Being, 62, 64, 98, 106, 119, 125, 129, 137, 150, 159, 162, 177

211

The Path of the Heart—*Cont.*
  Spectrum of Consciousness, 46, 58, 65,
    85, 98, 106, 118–120, 125, 137–138,
    150, 154, 159, 177, 179, 183, 185–186,
    191
theory of everything
  Lederman, L, 35, 183, *see also* quarks
  Quantum Mechanics, 13, 34, 181, 186,
    189, 192
three grades of SOUL
  Jewish Mysticism, 114
  nefesh, 114
  neshamah, 114
  offspring A, 114
  ruah, 114
TIAMET - 10
  evolution, 1, 3, 5–6, 9, 53, 55, 57, 62, 67,
    88, 91, 93, 95, 98–99, 104–107,
    112–114, 118, 122, 125, 127, 129, 135,
    137, 139–140, 147–148, 150–152, 154,
    159–160, 162, 165–168, 173, 177, 182,
    185–186, 188, 190–191
  star galaxy, 149, 152
  supernova, 149, 152
Tibetan
  Buddhism, 37, 43, 65, 102, 113, 120, 138,
    180, 193
  offspring I, 120
time
  linear time, 167
  non-linear, 20, 27, 29, 37, 42, 86, 184,
    187, 190–191
  NOW, 27
  sequential, 32
  simultaneous time, 101, 125, 128, 167,
    191
  timeless, 10–11, 14, 30, 40, 54, 64, 109,
    122, 125, 153, 166, 179, 182, 186
  timelessness, 28
TONAL
  Mayan, 40, 113, 118
  offspring F, 118
  sorcerer, 118, 133, 191
totems
  Childhood, 17, 28–29, 53, 92, 125, 147,
    153–154, 160
Trans-temporal
  mystical, 4, 15–16, 61–63, 66–67, 74, 91,
    98, 102, 108–109, 113, 121, 136–137,
    140, 146–147, 168, 185, 187
  time, 2, 4–5, 8, 10–15, 18–22, 24–32,
    40–42, 51, 61, 70, 77, 83–84, 90–93,
    101, 103, 107, 109–111, 117, 122–123,
    125–128, 132–134, 144–146, 148–152,
    155, 157, 159, 161–162, 165, 167–168,
    173, 179–180, 182–187, 189, 192–193
  timeless, 10–11, 14, 30, 40, 54, 109, 122,
    125, 153, 166, 179, 182, 186
  transcendental, 4, 58–59, 61–62, 64, 86,
    98, 112–113, 118–119, 157, 168, 188,
    190, 192
transcend ego
  evolution, 1, 3, 5–6, 9, 53, 55, 57, 62, 67,
    88, 91, 93, 95, 98–99, 104–107,
    112–114, 118, 122, 125, 127, 129, 135,
    137, 139–140, 147–148, 150–152, 154,
    159–160, 162, 165–168, 173, 177, 182,
    185–186, 188, 190–191
  Ex-centric ego, 156, 163
  Four Worlds, 62, 98, 106, 118, 121,
    124–125, 137, 150, 177
  spectrum of consciousness, 46, 58, 65,
    85, 98, 106, 118–120, 125, 137–138,
    150, 154, 159, 177, 179, 183, 185–186,
    191
transcendental realms, 59, 61
transformation, 23, 57–58, 60, 85, 88, 92, 103,
  106, 125, 128, 145, 156–157, 162, 186, 192

transpersonal
  awareness, 2, 4, 6, 20, 23, 27–28, 30, 39,
    53–58, 60, 85, 87, 90, 92, 98–99, 101,
    105–107, 109, 112–113, 117–122, 125,
    127, 130–131, 133, 137, 145, 147,
    155–157, 159, 162–163, 166, 168, 176,
    180, 184–186, 190
  consciousness, 2–6, 11, 13, 16, 19–21,
    23–25, 27, 29–31, 36–37, 39, 42–43, 46,
    51–58, 60, 62, 64–65, 67, 73, 79, 85–87,
    91–92, 95, 98–100, 103, 105–109,
    111–112, 117–120, 122–127, 129–130,
    135, 137–138, 140, 144–148, 150,
    153–157, 159–160, 162–167, 172–173,
    175, 177, 179–183, 185–189, 191–193
  Human, 2–3, 5–6, 9, 12, 21, 28–29,
    56–58, 64–65, 85, 98–99, 101, 147,
    157, 159–160, 162–163, 165, 173, 177,
    179, 181, 192
  Transpersonal Psychology, 58, 59, 62–64,
    85, 87, 94, 118, 162–168, 177, 192,
    194
  Transpersonal realms, 57, 59, 91, 118
  Transpersonal Scientist, 92, 145
  Transpersonal Self, 91–93
  TRANSPERSONAL SPECIESHOOD,
    65, 101, 152, 160, 162, 166
Tree of Life
  Jewish Mysticism, 114
  offspring A, 114
tripartite unity
  offspring A, 114
  trinity, 114–116
Twilight Zone
  Serling, R., 166
twin electron
  non-locality, 13, 15–16, 18, 20, 23–24,
    37, 123, 187
  Speed of Light, 15, 35

Uncertainty
  Chaos, 2, 4, 9, 19–20, 28, 34, 36–38, 40,
    42–43, 45–46, 52, 65, 84, 89, 101, 147,
    150, 156, 159, 163–164, 173, 177, 180,
    187
  complexity, 9, 20, 27, 34, 37, 42, 45–46,
    52, 147, 150, 157, 159
  indeterminism, 9, 34, 36
  Quantum Mechanics, 13, 15–16, 34, 36,
    181, 186, 189, 192
unconscious
  collective, 2–3, 5–6, 22, 42, 50–51, 53,
    59, 63–64, 85, 88, 91, 94, 108, 112,
    127–128, 168, 179, 181, 190
  see also Jung, C.G.
  personal, 4–6, 23, 29, 35, 50–52, 55,
    58–59, 61, 65, 70, 72–75, 79–80, 85,
    87–88, 90–92, 94, 104, 106, 110,
    113–114, 117–118, 127–128, 132–135,
    139, 150, 155, 159, 167–168, 185, 188,
    190
undivided, unbroken, seamless whole, 121
uni-cellular, 149–150
unidirectional arrow of time, *see also* linear
  arrow of time
unitary field, 15, 106
unity and diversity
  reconciliation, 3, 29, 40, 52, 65, 88, 163,
    167, 173
Universe, 2, 4, 6, 9–10, 12–21, 23–25, 28–29,
  31–32, 34–35, 40–43, 50, 61, 64–65, 85–86,
  88, 92–93, 106–107, 109–110, 112, 114,
  123, 126–127, 129–130, 132–135, 137,
  147–149, 151–153, 155, 159, 161, 163–166,
  168, 172, 177, 179–182, 186–187, 189,
  191–193
Unmanifest, 13, 15, 37, 47, 106, 135, 187

unpredictability - *see also* uncertainty
  chaos, 9, 20, 28, 34, 37–38, 40, 42–43,
    45–46, 52, 65, 79–80, 84, 89, 101, 147,
    156, 159, 163–164, 173, 177, 180, 187
  new paradigm, 9, 34, 37–39, 50, 59, 86,
    121, 138, 167
  non-linear, 20, 37, 42, 86, 184, 187,
    190–191
  uncertainty, 2, 4, 9, 28, 34, 36–38, 42–43,
    84, 147, 156, 159, 167, 173, 177, 192

Vedanta, Hinduism
  offspring H, 113, 118–119
vijnanamayakosha
  Hindu, 16, 20, 113, 119–120, 134, 186
  offspring H, 113, 118–119
Virgin Mary
  Alexandra, 135–136
  revelation, 116, 135, 177, 190
virtual reality
  Joe, 130–131
  Maya, 16, 110, 130, 186
Vishnu, 134
visionary
  mystic, 3, 14, 16–17, 19, 27, 41, 79, 93,
    99, 114, 117, 121, 125, 138–139, 187
  seer, 86–87, 117, 190
VOID, 10–11, 13–14, 109, 121, 192

WARNING SIGNS, 156
web of life, 10, 17, 20
whole, 4–5, 9, 15–16, 18, 20, 35–37, 52, 56,
  59, 64–65, 72, 86, 89, 101, 103, 105,
  108–110, 121, 124, 136–138, 148, 154, 163,
  168, 174, 176–177, 184, 187, 190
Wife, 59, 81, 89, 93, 145
Wilber, K., 62, 65–66
WISDOM age
  section seven, 171
  Vision, 8, 67, 90, 138, 146–147, 172
Wolf, F.
  dreaming universe, 126, 133
  Imaginal Realm, 23, 117, 184
  offspring E, 117
  Shaman, 4, 23, 62, 66, 117, 184, 191
Woolger, Jennifer & Roger, 52
Word made Flesh
  Catholicism, 114
  offspring B, 114, 116
World of 2070 A.D.
  section seven, 171–174
  Vision, 8, 67, 90, 138, 146–147, 172
World of Creation
  Four Worlds of mystic, 64, 108–110, 113,
    121–124
world of density
  manifest world, 64, 98, 105, 111–116, 118,
    121–124, 130, 132, 147, 150, 155, 185
  physicality, 2, 15, 41, 90, 109, 111, 124,
    126, 130–131, 135, 137, 148, 153, 157,
    159, 163–164, 186
World of Formation
  forms, 6, 11, 22–23, 28, 30–31, 38,
    40–42, 51, 77, 79, 85–86, 88, 105–110,
    118, 121–122, 124, 129, 135, 150–151,
    179, 188
  Four Worlds of mystic, 121–122
  ideas, 2–3, 13–15, 20, 22, 24, 26, 31, 37,
    39, 47, 50–52, 56, 64–71, 74–75,
    81–82, 85–87, 89, 94, 98–99, 103, 105,
    107–108, 110–112, 118, 121, 126,
    127–128, 131–136, 148, 166, 182,
    187–188, 191
  non-local, 13, 15, 16, 18, 20, 22–24, 37,
    123, 187

Index

patterns, 21, 23, 30, 37, 41, 51–52, 59, 61, 79–80, 84–85, 89, 110, 121–122, 131, 161, 179, 188
worlds of Emanation
    ALL THAT IS, 35, 43, 98, 109, 111–112, 114, 119, 121, 127–128, 144, 147, 161–163, 166, 179, 189, 191
    Four Worlds, 62, 98, 106, 108, 116, 118, 121, 124–125, 137, 150, 177
    SOURCE, 12, 15, 27, 38, 42, 64, 86–87, 99, 106, 108–109, 112–113, 115–116, 121–122, 125, 127, 134, 137, 148

worlds within worlds
    ALL THAT IS, 35, 43, 98, 109, 111–112, 114, 119, 121, 127–128, 144, 147, 161–163, 166, 179, 189, 191
    Four Worlds, 62, 98, 106, 108, 116, 118, 121, 124–125, 137, 150, 177
    SOURCE, 12, 15, 27, 38, 42, 64, 86–87, 99, 106, 108–109, 112–113, 115–116, 121–122, 125, 127, 134, 137, 148

Yaqui sorcerer
    Castaneda, C., 66, 132–133
    offspring F, 118
Yetzirah
    Four Worlds, offspring A, 114–116

Zen master Dogen, 36

Once Upon ASOUL
The Story Continues...

Joyce A. Kovelman, Ph.D. Anatomy, Ph.D. Psychology

NEW
Available
DECEMBER 1997

Once Upon ASOUL invites everyone who recognizes that there is more to Heaven and Earth than readily visible and who yearns to explore the deeper mysteries of life, to share the collective wisd of science, psychology, and spirituality as it integrates mind, body, and spirit into a satisfying wh ness.

Once Upon ASOUL speaks to an enduring, eternal core of Consciousness, a sacred essence that e of us carries deep within our being.

Once Upon ASOUL invites each of us to participate with Consciousness Itself, in order to explore sacred and deep mysteries beyond ordinary existence.

Transpersonal self is guardian of the environment and an instrument for peace... partner and co-c ator of our world.

"A vivid and compelling synthesis of science, psychology and spirituality."

— Marianne Williamson, Author of *Return to Love*, *A Woman's Worth*, and *Illuminata*

"Once Upon ASOUL is a thrilling journey through science, psychology, and the world of the spirit. No one can afford to neglect the information in this book, which could only have been written by a seasoned travele and very wise woman. For an introduction to the next millennium, I highly recommend this book."

— Larry Dossey, M.D., Author of *Healing Words*, *Recovering the Soul*, and *Space, Time & Medicine*

"Joyce Kovelman brings together the new physics, transpersonal psychology, the perennial philosophy, and mystical spirituality in a breathtaking synthesis. She places the human journey within the largest possible context and offers a powerful vision of our origins and unfolding future."

— Michael Washburn, Author of *The Ego* and *The Dynamic Ground Transpersonal Psychology Psychoanalytic Perspective*

Science, Psychology and The Realms of the Spirit	ISBN: 1-880396-52-1	**$24.95**
Credit cards, checks and money orders accepted.	Order No. JP9652-1	
Combination Price $32.25 — *A 15% SAVINGS!* Order No. 5253-C	Add 10% shipping, mimimum $4.	

JALMAR PRESS • 24426 S. Main Street, Suite 702, Carson, CA 90745
To order: **(800)662-9662** • (310)816-3085 • Fax (310)816-3092